The Mathematician's Brain

❖

The Mathematician's Brain

❖

David Ruelle

PRINCETON UNIVERSITY PRESS

PRINCETON AND OXFORD

Library of Congress Cataloging-in-Publication Data

Ruelle, David.
The mathematician's brain / David Ruelle.
p. cm.
Includes bibliographical references and index.
ISBN-13: 978-0-691-12982-2 (cl : acid-free paper)
ISBN-10: 0-691-12982-7 (cl : acid-free paper)
1. Mathematics—Philosophy. 2. Mathematicians—Psychology. I. Title.
QA8.4.R84   2007
510—dc22   2006049700

British Library Cataloging-in-Publication Data is available

This book has been composed in Palatino

Printed on acid-free paper. ∞

press.princeton.edu

Printed in the United States of America

1 3 5 7 9 10 8 6 4 2

❖ *Contents* ❖

v

CONTENTS

vi

❖ *Preface* ❖

ΑΓΕΩΜΕΤΡΗΤΟΣ ΜΗΔΕΙΣ ΕΙΣΙΤΩ

ACCORDING TO TRADITION, Plato put a sign at the entrance of the Academy in Athens: "Let none enter who is ignorant of mathematics." Today mathematics still is, in more ways than one, an essential preparation for those who want to understand the nature of things. But can one enter the world of mathematics without long and arid studies? Yes, one can to some extent, because what interests the curious and cultivated person (in older days called a philosopher) is not an extensive technical knowledge. Rather, the old-style philosopher (i.e., you and me) would like to see how the human mind, or we may say the mathematician's brain, comes to grips with mathematical reality.

My ambition is to present here a view of mathematics and mathematicians that will interest those without training in mathematics, as well as the many who are mathematically literate. I shall not attempt to follow majority views systematically. Rather, I shall try to present a coherent set of facts and opinions, each of which would be acceptable to a fair proportion of my mathematically active colleagues. In no way can I hope to make a complete presentation, but I shall exhibit a variety of aspects of the relation between mathematics and mathematicians. Some of these aspects will turn out to be less than admirable, and perhaps I should have omitted them, but I felt it more important to be truthful than politically correct. I may also be faulted for my emphasis on the formal and structural aspects of mathematics; these aspects, however, are likely to be of most interest to the reader of the present book.

Human communication is based on language. This method of communication is acquired and maintained by each of us through contact with other language users, against a background of human experiences. Human language is a vehicle of truth but also of error, deception, and nonsense. Its use, as in the present discussion, thus requires great prudence. One can improve the precision of language by explicit definition of the terms used. But this approach has its limitations: the definition of one term involves other terms, which should in turn be defined, and so on. Mathematics has found a way out of this infinite regression: it bypasses the use of definitions by postulating some logical relations (called *axioms*) between otherwise undefined mathematical terms. Using the mathematical terms introduced with the axioms, one can then define new terms and proceed to build mathematical theories. Mathematics need, not, in principle rely on a human language. It can use, instead, a formal presentation in which the validity of a deduction can be checked mechanically and without risk of error or deception.

Human language carries some concepts like *meaning* or *beauty*. These concepts are important to us but difficult to define in general. Perhaps one can hope that mathematical meaning and mathematical beauty will be more accessible to analysis than the general concepts. I shall spend a little bit of time on such questions.

The contrast is striking between the fallibility of the human mind and the infallibility of mathematical deduction, the deceptiveness of human language and the total precision of formal mathematics. Certainly this makes the study of mathematics a necessity for the philosopher, as was stressed by Plato. But while learning mathematics was, in Plato's view, an essential intellectual exercise, it was not the final aim. Many of us will concur: there are more things of interest to the philosopher (i.e., you and me) than the mathematical experience, however valuable that experience is.

This book was written for readers with all kinds of mathematical expertise (including minimal). Most of it is a nontechnical discussion of mathematics and mathematicians, but I have also inserted some pieces of real mathematics, easy and less easy. I urge the reader, whatever his or her mathematical background, to make an effort to understand the mathematical paragraphs or at least to read through them rather than jumping straight ahead to the other chapters.

Mathematics has many aspects, and those involving logic, algebra, and arithmetic are among the most difficult and technical. But some of the results obtained in those directions are very striking, are relatively easy to present, and have probably the greatest philosophical interest to the reader. I have thus largely emphasized these aspects. I should, however, say that my own fields of expertise lie in different areas: smooth dynamics and mathematical physics. The reader should thus not be astonished to find a chapter on mathematical physics, showing how mathematics opens to something else. This something else is what Galileo called the "great book of nature," which he spent his life studying. Most important, the great book of nature, Galileo said, *is written in mathematical language.*

The Mathematician's Brain

❖ 1 ❖
Scientific Thinking

$\mathbf{M}$Y DAILY WORK consists mostly of research in mathematical physics, and I have often wondered about the intellectual processes that constitute this activity. How does a problem arise? How does it get solved? What is the nature of scientific thinking? Many people have asked these sorts of questions. Their answers fill many books and come under many labels: epistemology, cognitive science, neurophysiology, history of science, and so on. I have read a number of these books and have been in part gratified, in part disappointed. Clearly the questions I was asking are very difficult, and it appears that they cannot be fully answered at this time. I have, however, come to the notion that my insight into the nature of scientific thinking could be usefully complemented by analyzing my own way of working and that of my professional colleagues.

The idea is that scientific thinking is best understood by studying the good practice of science and in fact by being a scientist immersed in research work. This does not mean that popular beliefs of the research community should be accepted uncritically. I have, for example, serious reservations with regard to the mathematical Platonism professed by many mathematicians. But asking professionals how they work seems a better starting point than ideological views of how they should function.

Of course, asking yourself how you function is introspection, and introspection is notoriously unreliable. This is a very serious issue, and it will require that we be constantly alert: what are good and what are bad questions you may ask yourself? A physicist knows that trying to learn about the nature of time by introspection is pointless. But the same physicist will be willing to explain how he or she tries to solve certain kinds of problems (and this is also introspection). The distinction between acceptable and unacceptable questions is in many cases obvious to a working scientist and is really at the heart of the so-called scientific method, which has required centuries to develop. I would thus refrain from saying that the distinction between good and

1

bad questions is always obvious, but I maintain that scientific training helps in making the distinction.

Enough for the moment about introspection. Let me state again that I have been led by curiosity about the intellectual processes of the scientist and in particular about my own work. As a result of my quest I have come to a certain number of views or ideas that I have first, naturally, discussed with colleagues.[1] Now I am putting these views and ideas in writing for a more general audience. Let me say right away that I have no final theory to propose. Rather, my main ambition is to give a detailed description of scientific thinking: it is a somewhat subtle and complex matter, and absolutely fascinating. To repeat: I shall discuss my views and ideas but avoid dogmatic assertions. Such assertions might give nonprofessionals the false impression that the relations between human intelligence and what we call *reality* have been clearly and finally elucidated. Also, a dogmatic attitude might encourage some professional colleagues to state as firm and final conclusions their own somewhat uncertain beliefs. We are in a domain where discussion is necessary and under way. But we have at this time informed opinions rather than certain knowledge.

After all these verbal precautions, let me state a conclusion that I find hard to escape: *the structure of human science is largely dependent on the special nature and organization of the human brain.* I am not at all suggesting here that an alien intelligent species might develop science with conclusions opposite to ours. Rather, I shall later argue that what our supposed alien intelligent species would understand (and be interested in) might be hard to translate into something that we would understand (and be interested in).

Here is another conclusion: *what we call the scientific method is a different thing in different disciplines.* This will hardly surprise those who have worked both in mathematics and in physics or in physics and in biology. The subject matter defines to some extent the rules of the game, which are different in different areas of science. Even different areas of mathematics (say, algebra and smooth dynamics) have a very different feel. I shall in what follows try to understand the *mathematician's brain.* This is not at all because I find mathematics more interesting than physics and biology. The point is that mathematics may be viewed as a production of the human mind limited only by the rules of pure logic. (This statement might have to be qualified later, but it is good enough for our present purposes.) Physics, by contrast, is also constrained by the physical reality of the world that surrounds us. (It may be difficult to define what we mean by physical reality, but it does very much constrain physical theory.) As for biology, it deals with a group of Earth-bound organisms that are all historically related: this is quite a serious constraint.

The two "conclusions" I have just proposed are of limited value because they are stated in such general and vague terms. What is interesting is to get into the details of how science is done and what it captures of the elusive nature of things. What I call the nature of things or the structure of reality is what science is about. That includes the logical structures studied by mathematics and the physical or biological structures of the world we live in. It would be counterproductive to try to define *reality* or *knowledge* at this point. But clearly there has been an immense progress in our knowledge of the nature of things over the past centuries or decades. I would go beyond that and claim a third conclusion: *what we call knowledge has changed with time.*

To explain what I mean, let me discuss the example of Isaac Newton.[2] His contributions to the creation of calculus, mechanics, and optics make him one of the greatest scientists of all time. But he has left many pages of notes telling us that he had other interests as well: he spent a lot of time doing alchemical manipulations and also trying to correlate history with the prophecies of the Old Testament.

Looking back at Newton's work, we can readily see which part of it we want to call science: his calculus, mechanics, and optics had tremendous later developments. His alchemy and his study of prophecies by contrast did not lead anywhere. The lack of success of alchemy can be understood from the way of thinking of alchemists, which involved relations between the metals and the planets and other concepts that we consider to be without rational or empirical justification. As to the esoteric use of the Scriptures to understand history, it continues to this day, but most scientists *know* that this is nonsense (and this opinion is supported by statistical studies).[3]

A modern scientist distinguishes readily between Newton's good science and his pseudoscientific endeavors. How is it that the same admirable mind that unveiled the secrets of celestial mechanics could completely go astray in other domains? The question is irritating because we see good science as honest and guided by reason while pseudoscience is often dishonest and intellectually off the track. But what track? What we see now as the well-marked path of science was at Newton's time an obscure track among other obscure tracks that probably led nowhere. The progress of science is not just that we have learned the solution of many problems but, perhaps more important, that we have changed the way we approach new problems.

We have thus gained new insight into what are good and bad questions and what are good and bad approaches to them. This change in perspective is a change in the nature of what we call *knowledge*. And this change of perspective gives a contemporary scientist, or an educated layman, some intellectual superiority over giants like Newton. By intellectual superiority I mean not just more knowledge and better methods but in fact a deeper grasp of the nature of things.

❖ 2 ❖

What Is Mathematics?

WHEN SPEAKING of mathematics, it is desirable to give examples. In this chapter the examples will be easy, but the reader should be warned against the natural tendency to accelerate through what appears to be technical stuff. On the contrary one should slow down! So, here we go.

Look at two triangles ABC and $A'B'C'$, and suppose that $|AB| = |A'B'|$. (This means that the length of the edge AB is the same as the length of the edge $A'B'$.) Suppose also that $|BC| = |B'C'|$ and that the angle at B in the triangle ABC is the same as the angle at B' in $A'B'C'$.

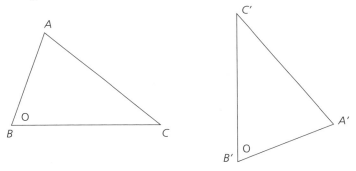

Having supposed all these things, it follows that the two triangles ABC and $A'B'C'$ are equal or, as one says, *congruent*. What this means is that if the two triangles are drawn on paper and you cut them out with scissors, you can move them around and superpose them exactly. (You may have to flip one of the two triangles recto-verso before putting it on top of the other.) Using the pieces of paper you can also make clear what you mean by *equal edges* (they can be superposed exactly) or *equal angles* (same thing).

If you have reasonable command of the English language and a modicum of visual intelligence, you will have understood the above considerations and quite likely found them mortally boring. Indeed, by the time you have taken in what is really meant,

5

these considerations will probably appear to you painfully trivial and obvious. Why then did people ever get excited about "theorems of geometry" such as the one we just discussed? For the hell of it let us state it again: *if the triangles ABC and A'B'C' are such that* $|AB|=|A'B'|$, $|BC|=|B'C'|$, *and the angle at B in ABC is equal to the angle at B' in A'B'C', then ABC and A'B'C' are congruent.* And it is also true that *if the triangles ABC and A'B'C' are such that* $|AB|=|A'B'|$, $|BC|=|B'C'|$, *and* $|CA|=|C'A'|$, *then ABC and A'B'C' are congruent.*

The fact is that from fairly obvious statements like this one it is possible, with impeccable logic, to derive more interesting results like the Pythagorean theorem[1]: *if the angle at B in the triangle ABC is a right angle,* then* $|AB|^2 + |BC|^2 = |AC|^2$.

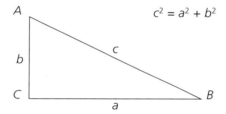

Actually, a proof of this result can be obtained by staring at the following figure:

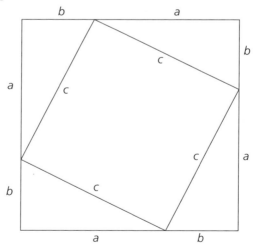

---

* You know what a right angle is, but if you insist on a definition, here is one: if the four angles of a quadrangle are equal, then they are right angles (and the quadrangle is a rectangle).

The big square has an area $(a + b)^2 = a^2 + 2ab + b^2$ and consists of a small square of area c^2 and four triangles with area $ab/2$ each, and hence, $a^2 + 2ab + b^2 = c^2 + 2ab$, that is, $a^2 + b^2 = c^2$.

The Pythagorean theorem is useful knowledge. It allows us, for instance, to produce a right angle if we have a piece of string. Here is how. We make marks on the string so that it is divided into twelve intervals of equal length (we may call this length a *cubit*). Then we use our string to make a triangle with sides 3, 4, and 5 cubits: the angle between the sides of length 3 and 4 cubits will be a right angle.

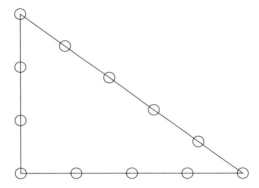

This is not quite obvious, but it follows from the Pythagorean theorem if you note that $3^2 + 4^2 = 9 + 16 = 25 = 5^2$. The ancient Greeks loved arguing, and they loved geometry because it gave them a chance to argue and to come to indisputable conclusions. Geometry, as Plato noted, is a matter of knowledge, not just of opinion. In Book VII of the *Republic*, he places geometry among the required studies for the philosophers who are going to rule his ideal city. In a very modern discussion, Plato remarks that geometry is practically useful but that the real importance of the subject lies elsewhere: "Geometry is knowledge of what always is. It draws the soul towards truth, and produces philosophical thought." Plato refers here to geometry in the plane and notes the lack of development (in his days) of three-dimensional geometry, regretting that "this difficult subject is little researched."[2]

Less than a century after Plato's *Republic*, Euclid's *Elements*[3] (ca. 300 BC) appears. The *Elements* gives a firmly logical presentation of geometry: a sequence of statements (called *theorems*) related by strict rules of deduction. One starts with some choice of statements that are assumed to be true (using modern language, we call them all *axioms*), and then the rules of deduction produce theorems that constitute geometry. Modern mathematicians are somewhat more fussy than Euclid in formulating axioms and proving theorems. In particular David Hilbert[4] showed that, to be really rigorous, one has to replace some of Euclid's intuitive thinking (based on looking at figures) by further axioms and more painful arguments. But the remarkable thing is that modern mathematics is done precisely in the way that Euclid presented geometry.

Let me say this again. Mathematics consists of statements— like the one about congruent triangles or the Pythagorean theorem—related by very strict rules of deduction. If you have the rules of deduction and some initial choice of statements assumed to be true (called *axioms*), then you are ready to derive many more true statements (called *theorems*). The rules of deduction constitute the logical machinery of mathematics, and the axioms comprise the basic properties of the objects you are interested in (in geometry these may be points, line segments, angles, etc.). There is some flexibility in selecting the rules of deduction, and many choices of axioms are possible. Once these have been decided you have all you need to do mathematics.

One terrible thing that could happen to you is if you reach a contradiction, that is, if you prove that some statement is both true and false. This is a serious concern because Kurt Gödel[5] has shown that it is not possible (in interesting cases) to prove that a system of axioms does not lead to contradictions. It is fair to say, however, that Gödel's theorem does not prevent mathematicians from sleeping. What I am trying to say is that most mathematicians are not distracted by Gödel from their usual routine: they don't expect a contradiction to pop up in their work. We may thus for the moment dismiss the issue of noncontradiction and look at "real" mathematics as it is usually done by mathematicians.

Mathematics as done by mathematicians is not just heaping up statements logically deduced from the axioms. Most such statements are rubbish, even if perfectly correct. A good mathematician will look for *interesting* results. These interesting results, or theorems, organize themselves into meaningful and natural structures, and one may say that the object of mathematics is to find and study these structures.

Here, however, one should be careful. I have followed the opinion of most mathematicians in saying that mathematics organizes itself in meaningful and natural structures. But why should that be the case? And, in fact, what does it mean? These are difficult questions. We shall look at them in the next chapter and later. Before that it is desirable to have a look at the role of language in mathematics.

When I say, "Consider the triangles ABC and $A'B'C'$, and assume that $|AB| = |A'B'|$, $|BC| = |B'C'|, \ldots$," I am using the English language. Sort of. The point here is not that mathematicians use bad English, but that they use English at all. Mathematical work is performed using a natural language (ancient Greek or English, for example), supplemented by technical symbols and jargon. We have said that mathematics consists of statements related by very strict rules of deduction, but now we see that the statements and deductions are presented in a natural language that does not obey very strict rules. Of course, there are grammatical rules, but they are so messy and fuzzy that translation by computer from one natural language to another is a difficult problem. Should the development of mathematics depend on a good structural understanding of natural languages? That would be quite disastrous.

The way out of this difficulty is to show that in principle we can dispense with a natural language like English. One can present mathematics as the manipulation of formal symbolic expressions ("formulas"), where the rules of manipulation are absolutely strict, with none of the fuzziness present in the natural languages. In other words one can *in principle* give a completely formalized presentation of mathematics. Why only in principle and not also in fact? Because formalized mathematics would be so cumbersome and untransparent as to be totally unmanageable in practice.

We may thus say that mathematics, as it is currently practiced by mathematicians, is a discussion (in natural language, plus formulas and jargon) about a formalized text, which remains unwritten. One argues quite convincingly that the formalized text could be written, but this is not done. Indeed, for interesting mathematics the formalized text would be excessively long, and also it would be quite unintelligible by a human mathematician.

There is thus in mathematical texts a perpetual tension: the need to be rigorous pushes towards a formalized style, while the need to be understandable pushes towards an informal exposition using the expressive possibilities of a natural language. There are a few tricks that make life simpler. An important one is the use of *definitions*: one replaces a complicated description (like that of a regular dodecahedron) by a simple phrase ("regular dodecahedron") or a complicated symbolic expression by a simple symbol. One may also introduce *abuses of language*: some controlled sloppiness that won't lead to trouble. Note that a completely formalized text could be checked for correctness mechanically, say, by a computer. But for an ordinary mathematical text, one has to depend on the somewhat fallible intelligence of a human mathematician.

Different mathematicians have different manners of expressing themselves. In the best cases, the style is clear, elegant, beautiful. As modern examples one may cite Jean-Pierre Serre's *Cours d'arithmétique*[6] and Steve Smale's review article "Differentiable dynamical systems."[7] The styles of these two pieces are very different, Serre being more formal, Smale less so. Smale uses hand-drawn figures to explain his mathematical constructions, a thing that Serre would not want to do. Although the styles are so different, most mathematicians would probably consider that Serre's book and Smale's article are both masterpieces of exposition.

❖ 3 ❖

The Erlangen Program

IF YOU DEFINE PRECISELY a set of axioms and rules of logical deduction, you have all you need to do mathematics. Mathematics, however, is not just a big pile of statements logically deduced from basic statements called *axioms*. Most mathematicians would say that good mathematics consists of those statements that are interesting, that good mathematics has meaning, and that it is organized in natural structures. It remains then to explain what are "interesting statements," "meaning," and "natural structures." Those concepts are not easy to define precisely, but mathematicians consider them important, and we have to try to understand them. There have been various attempts to define natural mathematical structures, and we shall focus on this concept. In fact, some mathematicians will insist that interesting or meaningful statements are those that relate to natural structures, and others will disagree, but we must postpone a discussion of this question until we have an idea of what mathematical structures are.

Felix Klein,[1] in his famous inaugural lecture in Erlangen (in 1872), proposed a concept of natural structures of geometry, now known as the Erlangen program. To discuss the views of Felix Klein we need to actually *do* some mathematics, in fact, some geometry. And the proper way to proceed would involve axioms, theorems, and proofs. But I do not want to assume that the reader has professional expertise in mathematics or wants to acquire it. I shall therefore use the approach of the Greeks before they formalized geometry in the manner of Euclid's *Elements*. I shall ask you to stare at figures and make simple deductions (or believe some statements I shall make). Think of yourself as a lover of philosophy in ancient Athens. As you come to the Academy to hear the discussions, you see a sign asking "nongeometers" (or nonmathematicians) to stay out. But you are not afraid. You enter.

To understand Klein's ideas, let us look first at the example of Euclidean geometry in the plane, which we discussed in chapter

11

2. We say that the plane is the *space* of Euclidean geometry, and we also have a notion of *congruence*. Two figures are congruent, or equal, if one of them can be moved to fit exactly on top of the other. The motion should be *rigid*, that is, it should not change distances between pairs of points. Rigid motions, i.e., congruences, are what characterizes Euclidean geometry. In Euclidean geometry we can speak of such concepts as (straight) lines, parallel lines, the middle point of a segment, the square, and so on. Euclidean geometry is very natural to us, but we shall see that there are other interesting geometries in the plane.

If we want to keep the concepts of straight lines and parallel lines but not those of distances between points or value of angles, we obtain *affine* geometry. Here, besides rigid motions we also allow stretching and shortening distances. Instead of congruences we have *affine transformations*. Note that a square, by a rigid motion, stays a square of the same size, oriented differently:

but by an affine transformation, a square can become any parallelogram:

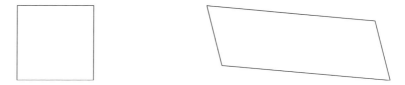

Affine geometry (of the plane) is defined by a space—the plane—and by the affine transformations. Let us mention in passing that the notion of the middle point of a segment makes sense in affine geometry, even though the notion of the length of a segment does not make sense. This is because we can say that segments of parallel lines are equal if they are intercepted by parallel lines:

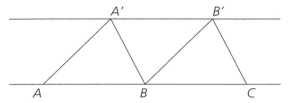

(if $A'A$ is parallel to $B'B$, and $A'B$ is parallel to $B'C$, then B is the midpoint of AC).

Another kind of geometry is *projective* geometry, which arises naturally from the study of perspective. Indeed, if you have a square table (left), its perspective drawing looks like this (right):

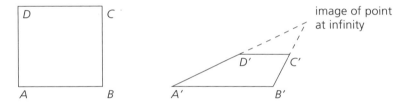

(the legs of the table have not been drawn). Note that the parallel sides of the table are no longer parallel in the picture. A natural idea here is to say that parallel lines intersect at a point at infinity. In the picture a point at infinity becomes an ordinary point of the plane.

In projective geometry we have a space called the projective plane, which consists of the ordinary points of the plane and the points at infinity. The rigid motions (or congruences) of Euclidean geometry are replaced by *projective transformations*, which move things around in a manner natural from the projective point of view: straight lines remain straight lines, but parallelism need not be preserved. If a figure is drawn on a table and you give a correct perspective rendering of this figure on a screen, you establish a projective transformation between the plane of the table and the plane of the screen. When a point P of the table is represented by a point P' on the screen, we may say that the projective transformation *sends P to P'*. As we have noted, a projective transformation may send a point at infinity to an ordinary point, and the converse also happens.

The midpoint of a segment is not a good concept for projective geometry, but the *cross-ratio* is. Take four points A, B, C, D on a line, and let a, b, c, d be their distances from a point O, with a + sign for points on the right of O and a − sign for those on the left. (The numbers a, b, c, d may thus be positive, negative, or 0.)

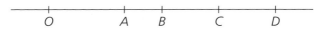

The quantity $(A, B; C, D)$ equal to

$$\frac{c - a}{d - a} : \frac{c - b}{d - b} = \frac{(c - a)(d - b)}{(d - a)(c - b)}$$

is called the cross-ratio of A, B, C, D. (It does not depend on the choice of O or what we called the right of O and the left of O.) If a projective transformation changes A, B, C, D to A', B', C', D', then $(A', B'; C', D') = (A, B; C, D)$. One can also define the cross-ratio of four lines PA, PB, PC, PD going through a point P:

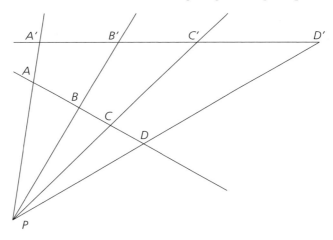

this is simply the cross-ratio of the points A, B, C, D as in the figure. (Using A', B', C', D' would give the same result.)

While the ideas we have just discussed go beyond Plato, he could have understood them. Let me now go briefly to something really different from Greek mathematics and use *complex numbers*. If you are not familiar with complex numbers, look at the note[2] before you proceed to the next paragraph. Plato might not have been happy with this next paragraph, and perhaps you won't be either. Read it anyway, but don't get stuck.

One may think of complex numbers as points in the complex plane. We define the *complex projective line* to consist of the points of the complex plane and a single extra *point at infinity*. Note that the complex projective line, which is really a plane, contains ordinary straight lines and circles. There are *complex projective transformations*, which move points around in the complex projective line. Specifically, the point (a complex number) z is sent (i.e., moved) to the point

$$\frac{pz + q}{rz + s}.$$

(We assume that p, q, r, s are complex numbers such that $ps - qr \neq 0$.) The complex projective transformations transform circles into circles (with the understanding that a straight line plus the point at infinity is considered to be a circle). One can define the cross-ratio of four points a, b, c, d (which are complex numbers) as

$$(a, b; c, d) = \frac{c - a}{d - a} : \frac{c - b}{d - b}.$$

It is in general a complex number, but when a, b, c, d are on a circle the cross-ratio is real (and the converse is true). It turns out that if a complex projective transformation maps a, b, c, d to a', b', c', d', then $(a, b; c, d) = (a', b'; c', d')$. In other words, complex projective transformations preserve the cross-ratio. (You are welcome to check this fact. Using the definitions I have just given it is a simple calculation.)

Let us now step back and see what we have. We have introduced various *geometries*, each with a *space* and a choice of *transformations*. In the cases we discussed, the space is a plane (with points at infinity possibly added). But the plane was just for easy visualization; other spaces (for example, three-dimensional space) could be used. In mathematical parlance, the words *space* and *set* are more or less equivalent, meaning a collection of "points" in the case of a space or "elements" in the case of a set. A transformation sends points in a space S to points in a space

S' (often S' is the same as S). The idea of Felix Klein is that a space and a collection of transformations define a geometry.

Introducing different geometries allows us to put some order in theorems. As an example, consider Pappus's theorem:

Suppose that two triangles ABC and A'B'C' are such that the lines AA', BB', CC' intersect in a single point P. Suppose that the lines AB and A'B' intersect at the point Q, the lines BC and B'C' at the point R, and the lines CA and C'A' at the point S. Then there is a straight line through Q, R, S.

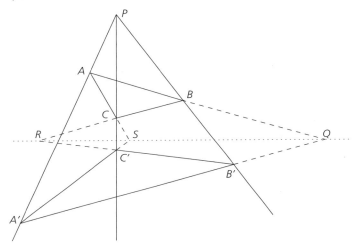

What kind of geometry does this correspond to? There are straight lines, no parallels, no circles. So, it is a good guess that Pappus's theorem belongs to projective geometry. Now, projective geometry is related to questions of perspective, and indeed Pappus's theorem can be understood in terms of perspective. Think of *ABC* and *A'B'C'* as being in fact triangles in three-dimensional space. We have assumed the existence of a point *P* , which is the top of a pyramid, of which *ABC* and *A'B'C'* are two plane sections. The planes containing *ABC* and *A'B'C'* must intersect in a line containing *Q, R, S,* and therefore there is a straight line through *Q, R, S,* as asserted by Pappus.

Perhaps it is at this point that you start feeling there is more to geometry than a legalistic certification of theorems. There are ideas—ideas that Plato could understand.

❖ 4 ❖

Mathematics and Ideologies

Now that we are able to distinguish between Euclidean, affine, and projective geometries, we may classify our knowledge accordingly. So, we have seen that Pappus's theorem belongs to projective geometry. But the Pythagorean theorem belongs to Euclidean geometry because it involves the concept of *length* of the sides of a triangle. Classification is a great source of satisfaction for scientists in general and mathematicians in particular. Classification is also useful: to understand a problem of Euclidean geometry, you will use one bag of tools including congruent triangles and the Pythagorean theorem. For a problem of projective geometry, you will use another bag of tools containing projective transformations and the fact that they preserve cross-ratios. A problem may be fairly easy if you use the right bag of tricks and become quite hard if you use the wrong one. Working mathematicians frequently experience this state of affairs and give due credit to Felix Klein for having uncovered this hidden mathematical reality: there are several different geometries, and it is useful to know where individual problems belong.

To convince you that the Erlangen program is a useful piece of mathematical ideology, I would like now to discuss a *difficult* problem. Here it is:

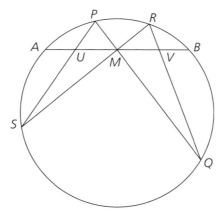

Butterfly Theorem
Draw a circle and a chord AB, with midpoint M. Then draw chords PQ
and RS through M as in the picture. Finally, let the chords PS and RQ
intersect AB at U and V, respectively. Claim: M is the midpoint of the
segment UV.

(Note that the *butterfly PSRQ* is usually an asymmetric quadrangle.) If you have some training in elementary geometry, I recom-
mend that you give this problem a good try before going on
(stop reading, take a sheet of paper, and start working).

Now let me explain that a professional mathematician would
not call this a very hard problem. With respect to difficulty, it
has nothing to do with *Fermat's last theorem*, which we shall discuss later. In fact, it looks like an easy question of elementary
Euclidean geometry. One immediately notices that the angles at
S and Q are equal. Then one tries to use standard results about
congruent triangles (like the one in chapter 2). Perhaps one
makes some constructions, drawing a perpendicular or a bi-
sector, . . . and one gets absolutely nowhere. Then one may start
having doubts. Is it really true that M is the middle of UV? (In
fact, yes, it is true.) The reasonable thing to do in this sort of
situation is to sleep on it. (I am a most reasonable person, so that
is exactly what I did after my colleague Ilan Vardi showed me
the problem and I could not readily solve it.) If you really want
to crack the problem, you can now do either of two things.

(i) Use brute force. In fact, for problems of elementary geometry one can always (as we shall see) introduce coordinates, write
equations for the lines that occur, and reduce the problem to
checking some algebra. This method is due to Descartes.[1] It is
effective but cumbersome. It is often long and inelegant, and
some mathematicians will say that it teaches you nothing: you
don't get a real understanding of the problem you have solved.

(ii) Find a clever idea so that the problem becomes easy. To
most mathematicians this is the preferred method.

Here the clever idea is to realize that the butterfly theorem belongs to projective rather than Euclidean geometry. True, the circle is a Euclidean object, but it also appears naturally in the geometry of the complex projective line. True, the notion of a midpoint
is Euclidean or affine, but this is a red herring: one might start
with $|AM| = \alpha |AB|$, where α is not necessarily $1/2$.

Let me now briefly outline a proof of the butterfly theorem. You may work out the details or be satisfied with the general idea, as you prefer. Consider the points A, B, P, R, ... to be complex numbers. Since A, B, P, R are on a circle, the cross-ratio $(A, B; P, R)$ is real. Taking the origin of the complex plane at S, we find that the points $A' = 1/A$, $B' = 1/B$, $P' = 1/P$, $R' = 1/R$ are on a straight line and

$$(A, B; P, R) = (A', B'; P', R').$$

This is also* the cross-ratio of the lines SA', SB'; SP', SR' or (by a reflection) of the lines SA, SB; SP, SR or (intersecting by AB) of the points A, B; U, M. Thus,

$$(A, B; P, R) = (A, B; U, M).$$

Replacing S by Q we find similarly

$$(A, B; P, R) = (A, B; M, V).$$

We have shown that

$$(A, B; U, M) = (A, B; M, V),$$

that is,

$$\frac{U - A}{M - A} : \frac{U - B}{M - B} = \frac{M - A}{V - A} : \frac{M - B}{V - B}.$$

If M is the midpoint of AB, that is, $M = (A + B)/2$, then the above equation simplifies and we have

$$(U - A)(V - A) = (U - B)(V - B),$$

or, expanding and regrouping,

$$(B - A)(U + V) = B^2 - A^2,$$

or, dividing by $B - A$,

$$U + V = A + B.$$

This shows that the midpoint of the segment UV is

---

* At this point we pass briefly from one-dimensional complex projective geometry to two-dimensional real geometry. By the way, instead of one-dimensional complex projective geometry, some mathematicians will prefer to use two-dimensional conformal geometry, but it is nearly the same thing.

19

$$\frac{U + V}{2} = \frac{A + B}{2} = M,$$

as announced.

So, we find that a difficult problem of geometry has a natural and elegant proof once we understand that the problem naturally belongs to projective rather than Euclidean geometry. This example, with many others, shows that there are natural structures in mathematics. These natural structures need not be easy to see. They are like the pure ideas or forms that Plato had imagined. The mathematician thus has access to the elegant world of natural structures, just as, in Plato's view, the philosopher can reach the luminous world of pure ideas. For Plato, in fact, a philosopher must also be a geometer. Today's mathematicians are thus the rightful descendants of the philosopher-geometers of ancient Athens. They have access to the same world of pure forms, eternal and serene, and share its beauty with the Gods. This sort of view of mathematics has come to be called *mathematical Platonism*. And in some form or other it remains popular with many mathematicians. Among other things it puts them above the level of common mortals. Mathematical Platonism can not be accepted uncritically, however, and we shall later dwell at length on this problem.

But at this point a shocking question should be discussed: how did our butterfly theorem find itself included in a list of "anti-Semitic problems"? The setting of the story is the Soviet Union, and the time is the 1970s and 1980s. As you may know, the Soviet Union was brilliant in many fields of science, in particular mathematics and theoretical physics. Scientific excellence was rewarded, and following the meanderings of the party line played a less dominant role in science than in other areas of Soviet life. Scientists then were to some extent sheltered from the prejudices of the ruling caste. But eventually the Soviet authorities, via the party committees of the universities, took steps to change this situation. In particular they limited the admission of Jews and certain other national minorities to major universities (in particular, Moscow University). This policy was not open and official but was implemented by selectively failing undesirable candidates at entrance examinations. Some details are given in papers by Anatoly Vershik and Alexander Shen,[2] where they provide a list of "murderous" problems used to flunk those who were not considered ethnically or politically correct. Proving the butterfly theorem is on the list of the "murderous problems," and you can see why: the natural way to attack the problem will probably lead you nowhere. Of course there is a relatively simple solution, and a seasoned mathematician will eventually find it. But think of a young person who takes an examination to enter the university and has to crack such a problem in limited time.

I have discussed Soviet politics with a number of Russian colleagues (now mostly in the West). One of them, explaining to me how he was selectively failed at the entrance examination to Moscow University in spite of giving the correct answers, appeared sad rather than angry. He is now a very successful research mathematician in the United States, but many others have had their lives distorted and ruined by the system. And, as one colleague remarked, "However tragic this problem of ethnic discrimination may have been in individual cases, it is only a marginal tragedy next to a much greater tragedy." Indeed think of the "excess mortality" of 16 million people in the camps of the Gulag, as officially recognized by the Soviet authorities. But even if you want to consider it as a marginal issue, the use of mathematics to implement ethnic and political discrimination is very disturbing to mathematicians. We thought of mathematics as living in a serene world of forms, beauty, pure ideas, and here it sits among other tools of repression.

Of course things have changed in Russia, and in the above-referenced article, Vershik mentions several university officials, formerly very active in the discrimination programs, who have suddenly turned into ardent democrats, organizing evenings of Jewish culture and the like. And some colleagues in the West seem eager to believe in this sudden mutation.

How did I drift from mathematics to this particular political discussion? I am not Russian, I am not Jewish, and the Soviet Union no longer exists. The plight of other groups is currently a more pressing question than that of Soviet Jews. So, should I not leave political ugliness aside and concern myself rather with the beauty of the Platonist world of forms? The fact is that while the political and moral aspects of science are not our main concern here, they cannot be totally ignored. I find scientists in general to be rather good company, but there is no question that some of them are bastards and some of them are frauds.[3] Now and then I am impressed by the moral strength of a colleague and now and then depressed by the moral weakness of one. Moral issues, one may say, are not part of science. But the exclusion and silencing of some scientists for extrascientific reasons may have far-reaching consequences. We shall later meet other examples of this unfortunate situation.

❖ 5 ❖

The Unity of Mathematics

STARTING FROM A LIST of axioms and rules of deduction, we have seen how geometry can be developed, proving one theorem after another. But there are more things in mathematics, than just geometry. For example, arithmetic: we start with the numbers 1, 2, 3, 4, ... called the (positive) *integers*. With integers one can make sums, $7 + 7 + 7 = 21$, and products, $7 \times 3 = 21$. One can define *primes*, 2, 3, 5, 7, ... , 137, ... (those are the integers that cannot be written as a product of two factors different from 1); we have just seen that 21 is not a prime. There are infinitely many primes, as Euclid already knew, but present-day mathematicians still have many questions about them.[1]

Some numbers that occur in geometry are not integers, for instance, the *fractions*, 1/2, 2/3, ... There are also *real* numbers that are not fractions, such as $\sqrt{2} = 1.41421\ldots$ or $\pi = 3.14159\ldots$ The number $\sqrt{2}$ is the diagonal of the square of side 1; it was known to Euclid (perhaps to Pythagoras) that $\sqrt{2}$ is not a fraction. The number π is the circumference of a circle of diameter 1; it is a modern result (from the eighteenth century) that π is not a fraction.

I might easily be carried away and start telling the saga of mathematics. How one proves a miraculous formula like[2]

$$1 + \frac{1}{4} + \frac{1}{9} + \frac{1}{16} + \cdots = \frac{\pi^2}{6}$$

and so on. But this is not my purpose here. What I have just said indicates two fundamental tendencies in the development of mathematics: diversification and unification.

It is clear how diversification arises: everyone can set up a new system of axioms and start deriving theorems, creating a new branch of mathematics. Of course one has to avoid systems of axioms that lead to contradictions, and to mathematicians, some systems of axioms will appear more interesting than others. But there are many branches of mathematics: *geometry,* which we discussed first; *arithmetic*, which deals with integers and related questions; and *analysis*, which is heir to the infinitesimal calculus of Newton and Leibniz.[3] And then there are more abstract topics called *set theory, topology, algebra*, and so on. So it would seem that mathematics disintegrates in front of our eyes into a dust of unrelated subjects.

But the subjects are not unrelated. For instance we have just seen how real numbers such as $\sqrt{2}$ or π appear in questions of geometry. In fact, there is a deep relation between Euclidean geometry and real numbers. Between Euclid and the nineteenth century the proper way to handle real numbers was through geometry: a real number was represented as the ratio of the length of two line segments. (This way of proceeding now appears clumsy to us, and this explains in part why we find Newton's mathematics rather painful to read). In the opposite direction, Descartes showed us how to do Euclidean geometry using real numbers. We choose rectangular axes Ox, Oy, and represent a point P_1 of the plane by its coordinates x_1, y_1 (which are real numbers), and similarly for P_2:

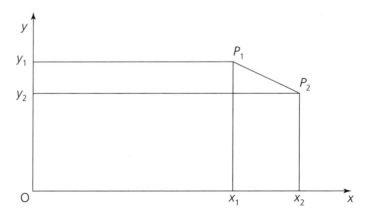

The length of the segment P_1P_2 is then $\sqrt{(x_2 - x_1)^2 + (y_2 - y_1)^2}$ by the Pythagorean theorem, and questions of geometry can be solved by formal manipulations of numbers (i.e., algebra).

In the above picture the point O has coordinates 0,0, and we can write $O = (0,0)$. Similarly, $P = (x, y)$ means that the point P has coordinates x, y. The circle of radius 1 with center O consists of those points $P = (x, y)$ such that

$$x^2 + y^2 - 1 = 0.$$

One says that $x^2 + y^2 = 1$ is the equation of the circle in question (drawn in the left of the following figure).

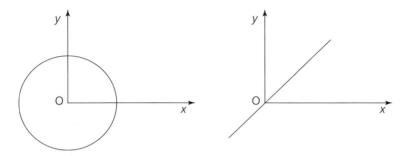

The slanted straight line through O in the right of the figure has the equation

$$x - y = 0.$$

The idea to represent curves by equations has been a very fruitful one, leading to what is called *algebraic geometry*.

Descartes' idea showed how to translate problems of geometry into problems about numbers. This was at a time when the modern theory of real numbers had not yet been developed. At present, however, we have simple and efficient axiomatic approaches to the real numbers. By contrast, the axiomatic description of Euclidean geometry (inherited from Euclid and made more rigorous by Hilbert) is a complicated mess. Currently, the efficient way to handle Euclidean geometry is to start from the axioms of real numbers, use the language of Descartes (Cartesian coordinates) to pass to geometry, establish as theorems a number of geometric *facts* (including the Euclid-Hilbert axioms), and then use these *facts* to derive theorems of geometry in the manner of Euclid.

Let me at this point digress on the role of axioms in the practice of mathematics: they are rather *unimportant*. This may come as a surprise after all the fuss we have made about defining mathematics in terms of axioms. What happens in mathematical practice is that one starts from a number of known *facts*: these may be axioms, or more often they are already proved theorems (like the Pythagorean theorem in Euclidean geometry). From these facts one then proceeds to derive new results.

The idea, then, is that one can construct one branch of mathematics using the axioms of another branch, supplemented by appropriate definitions. Doing this repeatedly, one can hope to present the whole of mathematics as a unified construction based on only a small number of axioms. This hope has been a central driving force for mathematics through the ninteenth and twentieth centuries. One can say that the hope has been fulfilled but not without crises and surprises. Names associated with the story are Georg Cantor,[4] David Hilbert, Kurt Gödel, Alan Turing,[5] Nicolas Bourbaki, and many more. We shall have the occasion to come back to parts of the story later but will now pause to look at the curious case of the French mathematician Nicolas Bourbaki.

For historical reasons France is a strongly centralized country. As a consequence, scientific research and teaching have often been under the control of a few powerful old people. And this control has been painful for the young bright scientists. After a while, however, the old tyrants die. The power is seized by the young bright scientists, who, in the mean time, have aged. After a while then they became old tyrants, and we are back to the old situation. One may lament this course of events, and its results are sometimes catastrophic for science. But sometimes the results are excellent. Why excellent? Because it frees the bright young people from time-consuming responsibilities and allows them to be scientifically productive. We have seen that for similar reasons Soviet science was of high quality in certain areas of research.

So, at the end of 1934, a few former students of the Ecole Normale Supérieure, the French mathematicians Henri Cartan (1904–), Claude Chevalley (1909–1984), Jean Delsarte (1903–1968), Jean Dieudonné (1906–1992), and André Weil (1906–1998), decided to fight the antiquated mathematics that surrounded them and write a treatise of *analysis*. Analysis contains things like *multiple integrals* and *Stokes formula*, which are of everyday use in theoretical physics. So the idea was to develop analysis in a fully rigorous manner, all the way to Stokes formula. This was going to be a collective effort, and the young revolutionaries decided to hide themselves under the pseudonym of Nicolas

Bourbaki (in jest after the somewhat forgotten nineteenth-century French general Charles Bourbaki).

Constructing analysis on a rigorous basis means we start from axioms. But not axioms of analysis! As we have seen, we want to build all of mathematics, including analysis, in a unified way from just a few axioms. And one century of mathematical mind searching has shown that a good idea is to start with the axioms of *set theory*. Set theory deals with collections of objects called *sets* (like the set {a, b, c} consisting of letters a, b, c). One can count the number of objects in a set and put together different sets (starting from {a, b, c} and {A, B, C} you get {a, b, c, A, B, C}). This looks tremendously uninteresting and unpromising. But because it is so simple, it is a good starting point for the whole of mathematics. In set theory you can analyze axioms and logical rules of deduction with the greatest clarity. How do you obtain the rest of mathematics from set theory? By counting the objects in a set, you get the integers 0, 1, 2, 3, ... From the integers you can define fractions and real numbers (using ideas of Richard Dedekind or Cantor). From the real numbers you obtain geometry (as we have seen). And so on. . . .

The outline of the work awaiting the young members of Bourbaki seemed rather clear, and they could hope that the job would not take too long. In fact many volumes on set theory, algebra, topology, and so on were written over the years by several generations of mathematicians (retirement age from Bourbaki was fixed at fifty). This work had considerable *normative* impact: notation and terminology were carefully discussed, and structural aspects of mathematics were investigated in great detail. In retrospect, it is clear that the rigorous, unifying, systematic ideology of Bourbaki has been an important component of twentieth-century mathematics. Even though not everybody liked it!

Of course the mathematicians who launched Bourbaki in the mid-'30s did not know where their enterprise would lead. But they were enthusiastic. And very sharp. Take for instance André Weil. To someone who said, "May I ask a stupid question?" he answered, "You just did." On the eve of World War II, André Weil decided that these nationalistic fights were not his problem, and he tried to escape the conflagration by taking refuge in Finland. There he was caught, barely escaped being executed as a spy, and was deported to England and then to France, where once again he barely escaped being executed but this time for trying to evade his military duties. Eventually he went to the United States, where he pursued a brilliant mathematical career, which eventually led to the *Weil conjectures*. The later proof of these conjectures by Grothendiek[6] and Pierre Deligne[7] was an important moment in twentieth century mathematics. Of course the political judgment of André Weil with respect to World War II may be questioned, but one must recognize the independence of mind that it shows. And this independence of mind served him well in his creative mathematical work, which shows a healthy disrespect for the achievements of his predecessors in the field of *algebraic geometry*.

Speaking of André Weil, we must mention his sister, Simone, who was the better known member of the Weil family in the European intellectual community. She was a philosopher and mystic, whose personal evolution took her from a Judaic background to Christianity. She wrote a number of influential books on her social and religious experiences. The prewar problems and wartime horrors hurt her deeply, and she died in 1943 (in England) from self-inflicted starvation.

And what about Bourbaki? He went from young and revolutionary to important and established to tyrannical and senile. The last two publications date from 1983 and 1998, and there will probably be no more. Bourbaki is dead.[8] The surviving members of Bourbaki (from the creative period) are now old mathematicians, for the most part covered with honors. Mathematics has integrated the ideas of Bourbaki, and then gone on.

❖ 6 ❖

A Glimpse into Algebraic Geometry
and Arithmetic

IF ONE WANTED to make a list of the ten greatest mathematicians of the twentieth century, the name of David Hilbert would be unavoidable. And one might want to add Kurt Gödel (although he was a logician more than a mathematician) and Henri Poincaré[1] (although he has been said to belong to the nineteenth century more than the twentieth). To go beyond these two or three names would be difficult, and different mathematicians might end up with rather different lists. We are still too close to the twentieth century to have a satisfactory perspective. Sometimes it will happen that a mathematician receives an important prize for the proof of a difficult theorem but that his reputation later fades. In other cases the work of a mathematician turns out in retrospect to have changed the course of mathematics, and his name will emerge as one of the very great names of science. One name that is certainly not fading at this time is that of Alexander Grothendieck. He was my colleague in France at the Institut des Hautes Etudes Scientifiques (IHES), and while I was not close to him, we were involved together in a series of events that led to his departure from the IHES and his exclusion from the mathematical community. He excluded himself. Or one might say that he rejected his former friends in the French mathematical community, and they rejected him. How this mutual rejection took place I shall later relate.

Before that, however, I want to say something of Grothendieck's mathematics, the last grand mathematical *œuvre* in the French language, the thousands of pages of the *Eléments de Géométrie Algébrique* and of the *Séminaire de Géométrie Algébrique*. Grothendieck started his career with problems in analysis, where he made contributions of lasting significance. But the great labor of his life was in algebraic geometry. This was highly technical work, but because of its magnitude, it has some visibility to nonspecialists, like the summit of a very high mountain that can be seen from afar even by those quite unable to climb it.

Algebraic geometry, as we saw, started with the description of curves in the plane by equations that we may write symbolically as

$$p(x, y) = 0.$$

A point $P = (x, y)$ of the curve has coordinates x, y, and in the examples that we considered earlier, the coordinates satisfied an equation $x^2 + y^2 - 1 = 0$ (circle) or $x - y = 0$ (straight line). Instead of $x^2 + y^2 - 1$ or $x - y$, we are now considering more generally a *polynomial* $p(x, y)$, that is, a sum of terms $ax^k y^\ell$ where x^k is the kth power of x, y^ℓ is the ℓth power of y, and the coefficient a is a real number. If $k + \ell$ can be only 0 or 1, we have a polynomial of the form

$$p(x, y) = a + bx + cy,$$

said to be of degree 1, and the curve described by $p(x, y) = 0$ is a *straight line*. If $k + \ell$ can be 0, 1, or 2, then we have a polynomial of degree 2,

$$p(x, y) = a + bx + cy + dx^2 + exy + fy^2,$$

and it corresponds to a conic. The curves called *conics* (or conic sections) were studied by later Greek geometers and include ellipses, hyperbolas, and parabolas.

Describing curves with the help of equations has the advantage that you can go back and forth between geometry and computation using polynomials. Take the following geometric fact: through five given points in the plane, there passes in general just one conic. The more precise theorem is the following: *if two conics have five points in common, then they have infinitely many points in common*. This geometric theorem is somewhat subtle but translates into a property of solutions of polynomial equations that makes more natural sense to a modern mathematician.[2] In general one may say that combining geometric language and intuition with the algebraic manipulation of equations has proved very rewarding.

The way a branch of mathematics develops is often guided by the subject studied, as if mathematicians were told, look at this, make that definition, and you will obtain a more beautiful, natural theory. This is what has happened in algebraic geometry: the way to develop the subject has been told to mathematicians, so to say, by the subject itself. For example, we have used points $P = (x, y)$, where the coordinates x, y were real numbers, but some theorems have a simpler formulation if you allow x, y to be complex numbers. For that reason, classical algebraic geometry largely uses complex rather than real numbers. This means that besides the real points of a curve one also considers complex points, and it is natural to introduce also *points at infinity*. Of course you will want to study not just curves in the plane but also curves (and surfaces) in three-dimensional space, and go to spaces of dimension greater than three. This forces you to consider a set of several equations, defining an *algebraic variety*, rather than just one equation. The way we just introduced them, algebraic varieties are defined by equations in the plane or some space of higher dimension. But it is possible to forget about the ambient space and to study varieties without reference to what is around them. This line of thinking was started by Riemann in the nineteenth century, and it led him to an intrinsic theory of complex algebraic curves.

The study of algebraic varieties is what algebraic geometry is about. It is a difficult and technical subject, yet it is possible to sketch in a general way how this subject has developed.

Let me come back to the remark that it is interesting to develop algebraic geometry with complex numbers rather than just real numbers. One can add, subtract, multiply, and divide real numbers in the usual way (dividing by 0 is not allowed), and this is expressed by saying that the real numbers form a *field*: the real field. Similarly the complex numbers form the complex field. And there are many other fields, some of which (called *finite fields*) have only finitely many elements. André Weil systematically developed algebraic geometry starting from an arbitrary field.

But why go from real or complex numbers to an arbitrary field? Why this compulsion to generalize? Let me answer by giving an example: instead of saying that $2 + 3 = 3 + 2$, or $11 + 2 = 2 + 11$, mathematicians like to state that $a + b = b + a$. It is just as simple, and being more general, it is also more useful. Stating things at the proper level of generality is an art. It is rewarded by obtaining a more natural and general theory and also, very importantly, by bringing answers to questions that could be stated but not answered in the less general theory.

At this point I would like to jump from algebraic geometry to something apparently quite different: arithmetic. Arithmetic asks, for instance, to find integers x, y, z such that

$$x^2 + y^2 = z^2.$$

One solution is $x = 3$, $y = 4$, $z = 5$, and the other solutions have been known since the Greeks. What if we replace the squares x^2, y^2, z^2 by nth powers with integer $n > 2$? Fermat's last theorem is the assertion that

$$x^n + y^n = z^n$$

has no solution with (positive) integers x, y, z if $n > 2$. In 1637, Pierre Fermat[3] thought that he had a proof of this assertion (later called *Fermat's last theorem*), but he was probably mistaken. A true proof was finally published in 1995 by Andrew Wiles.[4] This proof is extremely long and difficult, and one may well ask if it is worth spending so much effort to prove a result with zero practical interest. Actually, the main interest of Fermat's last theorem is that it is so hard to prove and yet can be stated so simply. Otherwise, it is just one consequence of the monumental development of arithmetic in the second half of the twentieth century.

Arithmetic is basically the study of integers, and a central problem of arithmetic is to solve polynomial equations (for instance, $p(x, y, z) = 0$, where we might have $p(x, y, z) = x^n + y^n - z^n$) in terms of integers (i.e., x, y, z are integers). Presented like this, arithmetic appears very similar to algebraic geometry: one tries to solve polynomial equations in terms of integers instead of complex numbers. Can one then unify algebraic geometry and arithmetic? Actually there are deep differences between the two subjects because the properties of integers are very different from those of complex numbers. In fact, if $p(z)$ is a polynomial in one variable z, there is always a complex value of z such that $p(z) = 0$. (This fact is known as the *fundamental theorem of algebra*.) But nothing like that is true for integers. To make a (very) long story short, it is possible to unify algebraic geometry and arithmetic, but it is at the cost of extensive foundational work. Algebraic geometry must be rebuilt on a (much) more general basis, and this was the great undertaking of Alexander Grothendieck.

At the time Grothendieck stepped into the game, a powerful idea had been introduced into algebraic geometry: instead of thinking of an algebraic variety as a set of points, one focussed on "nice" functions on the variety or parts of the variety. Specifically these functions are quotients of polynomials and make sense on parts of the variety where the denominators do not vanish. The nice functions just mentioned can be added, subtracted, or multiplied, but division is usually not possible (it does not give a nice function). The nice functions do not form a field but rather something called a *ring*. The integers (positive, negative, and zero) also form a ring. Grothendieck's idea was to start with arbitrary rings and see to what extent they might behave like the rings of nice functions in algebraic geometry and what conditions one should introduce so that the usual results of algebraic geometry continue to hold, at least in part.

Grothendieck's program was of daunting generality, magnitude, and difficulty. In hindsight we know how successful the enterprise has been, but it is humbling to think of the intellectual courage and force needed to get the project started and moving. We know that some of the greatest mathematical achievements of the late twentieth century are based on Grothendieck's vision: the proof of the Weil conjectures and the new understanding of arithmetic, which would make it possible to attack Fermat's last theorem. These applications of Grothendieck's ideas were, however, largely the work of other people and made after he had taken his leave from mathematics. In the next chapter I shall try to describe what had happened. But part of the story is certainly this: Grothendieck's passion was to develop new general ideas, to reveal grandiose mathematical landscapes. To accomplish this he needed to be forceful and clever. But cleverness was not an aim in itself. One may regret that he left behind him an unfinished construction, but adding detail carefully would not have interested him very much. Our great loss is not there but is that we don't know what other new avenues of knowledge he might have opened if he had not abandoned mathematics, or been abandoned by it.

❖ 7 ❖

A Trip to Nancy with
Alexander Grothendieck

THE IHES, where I came into contact with Grothendieck, is a small research institute for mathematics and theoretical physics located near Paris. It was created in the late 1950s by Léon Motchane as a privately funded institution to allow a few scientists to pursue their scientific work without other preoccupation. Motchane, an eccentric former businessman, had been born with the century and came originally from Saint Petersburg (he would never use the name Leningrad). When I met Motchane in 1964 he was a distinguished elderly gentleman who had done many things during his life. Before going into business, he had studied mathematics (with Paul Montel[1]). During World War II he was in the French Résistance against the Nazis.[2] He also spent part of his life in Africa. Asked what he had been doing there, I remember him answering, "Permit an old man to forget certain things. . . ." Part of the story behind the creation of the IHES was a carefully occulted romance between Léon Motchane and Annie Rolland: he was the first director and she was the secretary-general. Both retired in 1971, and then they got married, surprising many. Their dedication and the good advice that Motchane received from such people as the French mathematician Henri Cartan and the American physicist Robert Oppenheimer[3] ensured the original success of the Institut and the golden period of the 1960s and 1970s.

Just think that in the 1960s mathematics at IHES was represented by young René Thom[4] and young Alexander Grothendieck! While the scientific quality was stunning, the place appeared to me at the time totally unprestigious and unpretentious: nobody was an academician, and none of the scientists had gray hair. (Motchane had very distinguished white hair.) It may seem strange that I call the IHES *unprestigious* when René Thom had a Fields Medal (the most prestigious mathematical distinction at the time). The Fields Medal, however, played a less

34

important role then than it does now, particularly in France. Also, while Thom and Grothendieck had considerable intellectual ambitions, they were not pursuing prestige as such. Among memories I have of this period, let me record that once Grothendieck attended a seminar talk given by Thom (this was unusual). Grothendieck asked a question, to which Thom gave a somewhat fuzzy answer in his ordinary fashion. Grothendieck then retorted that this answer was the mistake made by all beginners! It would be too much to say that Thom enjoyed the criticism, but he could take it. All the scientists around were young and rather relaxed. The Institut at that time was new, outside of the French system, and there was no sense that we had to maintain any traditions. There was a great opportunity for everyone to follow his own intellectual path to the limit, and in different ways, that was what everyone did.

Alexander Grothendieck was born in Berlin in 1928.[5] His father had been a Russian revolutionary, who, not being a Bolshevik, left Russia after the triumph of Lenin. Grothendieck's father then fought in various European conflicts. He was with the Spanish Republicans when they were defeated by Franco, after which he took refuge in France. But as the war drew near, the French authorities put him in a camp. Later he would be handed over to the Germans and sent to his death in Auschwitz. In the 1960s an oil portrait of Grothendieck's father hung in the little office that Alexander Grothendieck occupied over the kitchen of the IHES. Alexander Grothendieck never saw much of his father and took the name of his mother (Hanka) Grothendieck. He spent his youth partly in Germany, partly in France, partly in freedom, partly in hiding, partly in camp.

At the end of the war, Grothendieck started studying mathematics at the University of Montpellier, where he stayed from 1945 to 1948. He was not satisfied with the level of rigor of what he was taught, and unaware of the concept of the Lebesgue integral (dating from 1902), he redeveloped measure theory on his own. In 1948 he went to Paris and was exposed to the modern mathematical world. He attended the lectures and *séminaire* of Henri Cartan and also met Jean-Pierre Serre, Claude Chevalley, and Jean Leray (1906–1998). These mathematicians were all to have a strong influence on the young Grothendieck. From 1949 to 1953, he was in Nancy (a much better place at the time than

Montpellier) and was recognized as a brilliant young mathematician for his work in functional analysis. After Nancy he spent a couple of years traveling (São Paulo, Kansas), and by the time he came to the IHES in 1958 his interests had shifted from functional analysis to algebraic geometry and arithmetic. For more than a decade he then worked like a titan on his *Eléments de Géométrie Algébrique*, and ran his Tuesday *Séminaire de Géométrie Algébrique* at the IHES, attended by a significant fraction of the French mathematical elite. Apart from Tuesdays when he came to the Institut, he worked at home. He worked a lot. His exceptional working capacity was combined with the two key qualities needed to be a creative mathematician: technical reliability and imagination. Besides that, the subject on which he was now working, combining algebraic geometry and arithmetic, was of proper size for his genius. But other circumstances also were favorable. He had no teaching or administrative duties. He spent little time studying the papers of other mathematicians: he had the ideas explained to him by colleagues. And a massive final asset was Jean Dieudonné.[6] Dieudonné, a high-class mathematician and a member of Bourbaki, also had enormous working capacity. He decided to act as Grothendieck's scientific secretary: he was able to understand his ideas and to write them neatly as mathematical text. These were the circumstances under which the miracle that was Grothendieck's contribution to mathematics took place.

Grothendieck's daily language was French and he used French for his mathematics, but his native tongue was German. Grothendieck had his father's portrait in his office, but he did not attach importance to his half-Jewish ancestry and he used his mother's name. The father fought in revolutionary wars, but the son was a firm antimilitarist. Alexander Grothendieck lived in France, but he chose to remain stateless (until 1980). He was a central figure of the French mathematical community, but he had the wrong pedigree: he was not a former student of the Ecole Normale. These discrepancies may have subtle explanations, but they point to a very plain fact: Grothendieck made deliberate choices about who he was. He could also change his mind after due reflection. Being a mathematician was his choice. But being a member of an exclusive elite of Paris mathematicians is something that happened to him, and he would later be quite upset that he had been trapped in this situation.

My scientific interests were rather different from Grothendieck's interests, but I knew him reasonably well as a colleague between 1964 and 1970. He was a rather handsome man, with a completely shaven head. He had obvious personal courage: he didn't cop out of difficult situations. Of this there would be ample evidence in confrontations with Motchane. He was actually less aggressive than many and always ready to discuss, but he would not accept arguments that he found unconvincing. He had charm, certainly, and moral questions like antimilitarism were important to him. At times, however, he could be insensitive and even brutal. If I were to compare him with other mathematicians, I would say he had a richer personality than most,[7] and also some of the intellectual rigidity that is frequent with them. It suited me that I was not close to him. But I must admit sympathy for the man, not just admiration for the mathematician. People with moral preoccupations, and courage, are uncommon. And scientists do not rate better than average in this respect.

Intellectually the IHES was a fantastic place when I arrived there in 1964. But there were also difficulties. One of the first things I heard (from René Thom) was that the paychecks of the professors did not always arrive on time. Later the IHES had to sell its assets to survive. (So I had to borrow money to buy the apartment I rented or get out.) The professors at the time (René Thom, Louis Michel,[8] Alexander Grothendieck, and myself) fully acknowledged the enormous dedication of Motchane, but we became concerned about his increasingly erratic behavior and about the problem of his successor. In 1969 we had a number of private meetings, and finally (in Louis Michel's dining room) we wrote a letter to Motchane asking him formally for a meeting of the scientific committee of the IHES to discuss the situation. This was not taken well, and things quickly turned sour. Motchane counterattacked, manipulating facts without much regard for the truth, threatening to close the Institut, and so on. On one occasion he presented us with an excerpt of the report of a recent committee meeting in which we supposedly had reappointed him as director for four years. Nobody remembered the decision, and we never saw the report from which this had been "excerpted." Motchane also tried to arrange his own succession without referring to the scientific committee (this was against the statutes of the IHES). Colleagues from outside wrote to us that they had been approached by Motchane, who had asked them not to tell us. They told us anyway but asked that we not mention this to Motchane. Looking more carefully at the statutes of the IHES, due to Motchane and at first sight very generous to the professors, we came to realize that they included no provision to force the director to do anything he didn't want to do. At one point during this confused period, Grothendieck found that the IHES had received military money and said that he would have to resign if this continued. But apparently the military money was coming to an end, and this closed the incident.

After discussion with Michel and Thom, Grothendieck and I went on February 20, 1970, to Nancy to meet with the president of the board of trustees. This was a largely meaningless trip, but it gave me a chance to spend a few hours on the train with Grothendieck. We talked about theoretical physics. He asked careful questions and discussed what I had to say with great attention. Around that time he also informed himself about biology with the help of his friend Mircea Dumitrescu. As do many people reaching the age of forty and with the destabilizing influence of the events of May 1968 in France, it is clear that Grothendieck was reconsidering the orientation of his life. He would not simply continue working as before on the foundations of algebraic geometry. He even claimed that he was leaving mathematics. This did not prevent him from doing excellent work after that, in spite of unfavorable conditions.

Let us go back to the IHES at the beginning of 1970. The atmosphere was deteriorating progressively, but we thought that things would improve in the end. I remember coming across Grothendieck in the Metro (in Massy-Palaiseau), and he told me, "We are all upset now with Motchane, but you will see that in a couple of years we shall laugh about the whole affair" (*on en rigolera*). This was, however, not to be. In discussions with Motchane, Grothendieck tended to be more outspoken than the rest of us. As a consequence Motchane obviously thought that he was our "leader" and taunted him. At some point Grothendieck seems to have decided that the farce had lasted long enough, and at one of our meetings with Motchane, he called him a bloody liar.[9] (I do not remember the precise reason for the accusation.) Following this incident, Motchane declared that the IHES again received military money, and Grothendieck resigned.

After that I no longer saw Grothendieck. He was involved with a small antinuclear group, he traveled, and he made a number of attempts to get a reasonable academic position in France. These attempts were largely unsuccessful: he had to be content with a teaching position in the provincial university at Montpellier, where he had been a student in 1945. In 1981 he speaks of a "bash in the face" (*un coup de poing en pleine gueule*) when his candidacy for a professor position was rejected by a committee comprising three of his former students. While mostly cut out from the research community, he had bouts of frantic mathematical activity, writing hundreds of pages that have been privately circulated, but only partly published.

Between 1983 and 1986, Grothendieck worked on the long text of *Récoltes et semailles* (Harvests and Sowings), more than 1,500 pages of reflections on life and mathematics. This is a very diverse text, parts of which read like Oscar Wilde's *De profundis*, while other parts contain paranoid attacks against former students and friends accused of having betrayed his scientific œuvre and message. These attacks are often too personal and private for comfort. They are no doubt partly unjust, partly true. *Récoltes et semailles* was privately circulated, but Grothendieck was unsuccessful at getting the text published as a book.[10] Parts of the text have strange beauty and depth. And it will remain a basic document for the understanding of an important period in the history of mathematics.

In 1988 Grothendieck turned sixty and took early retirement from Montpellier. This same year he "won" half of an important mathematical prize,[11] which he refused. He also was presented with a three-volume Festschrift,[12] about which he said that he thanked those who did not contribute to it.

In 1990 Grothendieck sent a letter announcing a prophetic book that he would write and publish in the same year. He said the 250 people who received the letter should prepare for a great mission ordered by God. But the announced book never appeared, and Grothendieck is now silent.

Since his retirement, Grothendieck has lived more and more as a recluse. He has been attracted to Buddhist ideas and follows an extreme vegetarian diet. His current whereabouts are not publicly known. I asked one of the last friends with whom Grothendieck had contact (in 2000) about how he was doing, and the answer was simply, "He meditates."

It may be hard to believe that a mathematician of Grothendieck's caliber could not find an adequate academic position in France after he left the IHES. I am convinced that if Grothendieck had been a former student of the Ecole Normale and if he had been part of the system, a position commensurate with his mathematical achievements would have been found for him. Let me digress on this a minute. When Pierre-Gilles de Gennes[13] got his physics Nobel Prize in 1991, there was an official ceremony in his honor at the Sorbonne, with speeches by de Gennes himself and the minister (Lionel Jospin). On this occasion de Gennes branded corporatism as the plague of French science. Certainly this applies to physics, and even more to mathematics. What this means is that it is all important whether you are from the Ecole Normale or the Ecole Polytechnique, in whose lab you were accepted, whether you are at the CNRS, the academy, a suitable political party, and so on. If you are in one of these groups, they will help you and you have to help them. In the case of Grothendieck, he was nothing (not even having at the time French or any other citizenship). He was nobody's responsibility; he was just an embarrassment.

Understandably, some people would like to blame Grothendieck's exclusion entirely on Grothendieck himself: he went crazy and left mathematics. But this does not fit with the known facts and their chronology. Something shameful has taken place. And the disposal of Grothendieck will remain a disgrace in the history of twentieth-century mathematics.

Structures

From what we have seen, mathematics appears to have a dual nature. On the one hand, it can be developed using a formal language, strict rules of deduction, and a system of axioms. All the theorems can then be obtained and checked mechanically. We may call this the *formal* aspect of mathematics. On the other hand, the practice of mathematics is based on ideas, like Klein's idea of different geometries. This may be called the *conceptual* (or *structural*) aspect.

An example of structural considerations arose in chapter 4 when we discussed the butterfly theorem. There we saw how important it is to know in what kind of geometry a theorem belongs when you want to prove it. But the concept of projective geometry is not explicit in the axioms that are currently used for the foundation of mathematics. In what sense is projective geometry present in the axioms of set theory? What are the structures that give a sense to mathematics? In what sense is the statue present in a block of stone before the chisel of the sculptor brings it out?

Before we discuss structures, it is a good idea to look a bit more at sets, which play such a fundamental role in modern mathematics. Let me review here some basic intuition, notation, and terminology. The *set* $S = \{a, b, c\}$ is a collection of objects a, b, c, called *elements* of S. The order in which the elements are listed does not matter. One writes $a \in S$ to express that a is an element of S. The sets $\{a\}$ and $\{b, c\}$ are *subsets* of $\{a, b, c\}$. The set $\{a, b, c\}$ is finite (it contains three elements), but sets may also be infinite. For instance, the set $\{0, 1, 2, 3, \ldots\}$ of natural integers or the set of points on a circle are infinite sets. Given sets S and T, suppose that for each element x of S, a (unique) element $f(x)$ of T is given. We then say that f is a *map* from S to T. One also says that f is a *function* defined on S and with values in T. For instance, one can define a map f from the set $\{0, 1, 2, 3, \ldots\}$ of natural integers to itself such that $f(x) = 2x$. Other maps (or functions) on the natural integers with values in the natural integers are given by

$$f(x) = xx = x^2 \text{ or } f(x) = x \cdots x = x^n.$$

The general concept of a function (or map) emerged slowly in the history of mathematics, but it is central to our current understanding of mathematical structures.[1]

Repeatedly, mathematicians have tried to define with precision and generality the structures that they use. Klein's Erlangen program is a step in that direction. The structures considered by Klein were geometries, each associated with a family of maps: congruences (for Euclidean geometry), affine transformations (for affine geometry), projective transformations, and so on. The very ideological Bourbaki has a definition of structures based on sets.

Let me try to give an informal view of Bourbaki's idea. Suppose that we want to compare objects with different sizes. We write $a \le b$ to mean that the object a is less than or equal to b. (Some conditions should be satisfied; for example, if $a \le b$ and $b \le c$, then $a \le c$.) We want thus to define an *order* structure ($\le$ is called an order). For this we need a set S of objects $a, b, \ldots$ that we shall compare. Then we can also introduce another set T consisting of pairs of elements a, b of S: those pairs for which $a \le b$. (We may have to consider other sets as well, to impose the condition that if $a \le b$ and $b \le c$, then $a \le c, \ldots$) In brief we consider several sets $S, T, \ldots$ in a certain relation (T consists of pairs of elements of S), and this defines an *order structure* on the set S. Other structures are similarly defined on a set S by giving each time various sets standing in a particular relation to S. Suppose, for instance, that the set S has a structure that allows its elements to be added; that is, for any two elements a, b, there is a third element c for which we can write $a + b = c$. Defining the structure of interest on S will thus involve consideration of a new set T of triplets of elements of S: those triplets (a, b, c) for which $a + b = c$. Mathematics textbooks contain the definitions of many structures with names like *group structure*, *Hausdorff topology*, and so on. These structures are the conceptual building blocks of algebra, topology, and modern mathematics in general.

Let the set S have an order relation, and let the set S' also have an order relation. Suppose that we have a way to associate with each element $a, b, \ldots$ of S an element $a', b', \ldots$ of S'. In mathematical language we have a map from S to S' sending $a, b, \ldots$ to $a', b', \ldots$. Suppose that if $a \leq b$, then $a' \leq b'$; that is, the map preserves the order. Let us use an arrow to denote the order-preserving map from S to S':

$$S \to S'.$$

More generally one likes to write $S \to S'$ to denote passing from a set with a certain structure to a set with similar structure, while preserving the structure. In technical jargon the arrow is said to represent a *morphism*. (So, if S and S' have a structure where elements can be added and the morphism $S \to S'$ sends elements $a, b, c, \ldots$ of S to elements $a', b', c', \ldots$ of S', then $a + b = c$ should imply $a' + b' = c'$.) If we consider sets with no added structure, the morphisms $S \to S$ are just all maps from S to S'.

A new development is to consider sets with a certain type of structure, together with the corresponding morphisms: this is called a *category*. (So there is a category of sets, where the morphisms are maps; a category of ordered sets, where the morphisms are order-preserving maps; a category of groups, and so on.) In this way of looking at things, it is useful to be able to map the objects of one category to those of another category while preserving the morphisms. When this is the case, one says that one has a *functor* from one category to another. Categories and functors were introduced around 1950 (by Samuel Eilenberg and Saunders Mac Lane[2]) and quickly developed into important conceptual tools in topology and algebra. Categories and functors may be viewed as the ideological backbone of an important part of late-twentieth-century mathematics, constantly used by some mathematicians like Grothendieck.

To summarize, one might say that in the ideological background of important areas of mathematics at the end of the twentieth century, there is a constant preoccupation with structures and their relations. Some questions will automatically be asked, and some constructions will automatically be attempted. To some extent, then, an answer has been given to the question of finding the conceptual building blocks of mathematics. The answer is in terms of structures, morphisms, and perhaps categories, functors, and related concepts. And the quality of this answer can be gauged by the wealth of the results obtained.

I must at this point correct a false impression that I may have given above: that current mathematical thinking is dominated by categories, functors, and the like. In fact, large and important parts of mathematics have little use for these concepts. One may just say that there is a general striving to clarify conceptual aspects and not just do brutish calculations. But structural considerations may be rather minimal. To give an example of a different style of mathematics, I want to mention the work of Paul Erdös[3] (the name is Hungarian, and the s at the end is pronounced "sh"). Erdös was a very unusual mathematician who kept traveling from place to place, without being attached to a fixed institution. His legacy to mathematics is diverse and important. He had the beautiful idea of The Book "in which God maintains the perfect proofs for mathematical theorems." (Incidentally, Erdös did not believe in God, whom he called the Supreme Fascist). Under the direct influence of Erdös, a fascinating approximation to The Book has been written, called *Proofs from The Book*.[4] This is reasonably easy and quite delightful to read, and gives a decidedly non-Bourbakist view of mathematics. It is not that structural considerations are absent, but they stay in the background. Paul Erdös was the type of mathematician called a problem solver, quite different from a theory builder like André Weil or Alexander Grothendieck. A good problem solver must also be a conceptual mathematician, with a good intuitive grasp of structures. But structures remain tools for the problem solver, instead of the main object of study.

The current conceptualization of mathematics continues the efforts of earlier periods and will no doubt be extended in the future. One might say that the philosophical quest for the fundamental structures of mathematics has been successful, in the sense that it has produced concepts that are uncannily efficient at producing new results and solving old problems. The fact that we have an efficient conceptualization of mathematics shows that this reflects a certain mathematical reality, even if this reality is quite invisible in the formal listing of the axioms of set theory.

The view that I have just presented comes close to what is called *mathematical Platonism*. In the *Republic*,[5] Plato speaks of a world of pure ideas to which the philosopher has access, while his less fortunate contemporaries, chained in an obscure cave, can only see fleeting shadows. The structures of mathematics (taking structure in a wide sense) are like the pure ideas of Plato, and the mathematician-philosopher has access to them, while his less fortunate contemporaries remain chained in non-mathematical obscurity. If we think of the structures of mathematics as statues, the mathematician does not chisel them out of a block of stone according to some random fantasy. Nay! The statues belong to the world of the Gods, and it is the mathematician's noble task to unveil them and reveal them in their eternal beauty.

You can probably see why the Platonist view appeals to many mathematicians, as different as Bourbaki and Erdös. I think, however, that it is partly misleading in that it ignores one essential fact: what we call mathematics is mathematics studied by the human mind or brain. The consideration of the mind may be irrelevant when we discuss the *formal* aspects of mathematics but not when we discuss *conceptual* aspects. Mathematical concepts indeed are a production of the human mind and may reflect its idiosyncrasies.[6]

Starting in the next chapter, I shall be concerned with the relation between the human mind (or brain) and this extraordinarily nonhuman thing that we call reality, in particular, mathematical reality. Having learned something about how our brain works, we shall be in a better position to address the big question: *how natural are the concepts and structures of mathematics?*

❖ 9 ❖

The Computer and the Brain

O NE OF THE MOST powerful and versatile scientific minds of the twentieth century was that of John von Neumann,[1] who made fundamental contributions in pure mathematics, physics, economics, and the development of the digital computer. *The Computer and the Brain*[2] was von Neumann's last book, and he wrote it while cancer was destroying his body. Published in 1958 after von Neumann's death, the book contains a fascinating comparison of the structure and functioning of the digital computer and the human brain.

But is such a comparison allowed? Is it not sacrilegious to compare the human mind, this noblest of all things, and the computer, a mere machine? Scientists are notoriously careless about sacrilege. Here we may note that the computer and the brain are both information-processing devices. This entails some similarities, like the need for a memory to store information. A comparison between the two devices is thus in order. And the comparison shows, as one might have expected, that the computer and the brain are very different, in many respects. Significantly, it appears that the functioning of the human brain has a number of peculiarities that are not shared with the computer and have thus no logical necessity. These peculiarities, and indeed shortcomings, can be expected to influence the way humans do mathematics, as I shall argue later.

But first let me make a point-by-point comparison between the computer and the brain, in the spirit of von Neumann (with some updating and different intentions). I shall start with a point "zero," as mathematicians often do, to set this point apart from the rest.

0. THE PRINCIPLES OF CONSTRUCTION OF THE COMPUTER AND THE BRAIN ARE DIFFERENT

The computer is a human invention. It processes and stores information in digital form (bits), and the way the information

46

should be processed and stored is defined in programs. Very different programs (software) can be put on the same machine (hardware), and this makes the computer extremely flexible and versatile.

The brain is a result of biological evolution. The genes in the egg cell of an animal contain some kind of blueprint for its nervous system (among other things). By trial and error (i.e., mutation and selection) the blueprint has been improved over eons. Improvement here means that, progressively, a nervous system is produced that gives a better chance of survival and reproduction to its possessor, under prevailing circumstances. The nervous system gets you to move away from harmful stuff, catch edible stuff, and decide action on the basis of sensory inputs. Over the last million years or two there was an explosive development of the central nervous system of our hominid ancestors. Finally, our species developed complex language, symbolic thinking, and a written tradition. And as a result, the human brain has become a flexible and versatile device, capable of solving relatively difficult questions (like "what are the prime factors of 169?") that a computer program could also solve but not an ape. (Of course there are also problems, like climbing up a tree, that an ape can solve better than a computer or a human!)

Speaking of evolution, let me stress that we have much better mathematical techniques than Euclid or Archimedes[3] had in their times. But we cannot claim that we are more intelligent than they were. This reflects the fact that our cultural evolution is much faster than biological evolution. As to the evolution of computers, it is extremely fast, both in terms of hardware (speed and memory size) and software (complexity and power of the programs they support). As a consequence, computers are progressively mastering difficult tasks such as playing chess or translating from one natural language to another. Let me make here a personal remark. I must admit that I am somewhat frightened by the rapid, apparently limitless evolution of computers. I see no reason why they could not overtake our cultural evolution and become, in particular, better mathematicians than we are. When this happens, I feel that life will have become for us somewhat less interesting and somewhat less worth living. Our world has seen the era of great Gothic cathedrals come to an end. And the era of great human mathematics may also come to its end. For the time being, however, mathematics goes on, and life goes on, and so we may proceed with our comparison of the computer and the brain.

1. THE BRAIN IS SLOW, AND ITS ARCHITECTURE IS HIGHLY PARALLEL

A computer operates by discrete units of time, or *cycles*, measured by a *clock*. Some new operation is effected at each cycle. Currently, the clock of your PC may work at a frequency of 1 Gigaherz; that is, a cycle is one nanosecond (= one billionth of a second). By contrast a characteristic time for a change in the nervous system would be at least one millisecond (= one thousandth of a second). But times of the order of 100 milliseconds can easily occur because the speed of propagation of the nervous influx is from 1 to 100 meters per second. So, what is instantaneous for the human brain is many millions of times slower than what your PC would call quick.

The high speed of computers is well suited to repetitive tasks, where each stage provides an updated input for the next stage. By contrast, the brain typically processes information in one sweep, using its massively parallel structure. This means, for instance, that the optic nerve carries information from different areas of the retina, in parallel, to different areas of the brain. In fact, a distorted image of the retina (and thus of the world in front of you) is projected on the visual cortex at the back of the brain, and different aspects of the image (color, orientation, etc.) are processed simultaneously. Parallelism has also been introduced in the structure of some computers, so-called special purpose machines, but this is not comparable with what is present in the more than 10^{10} neurons of our brain.

There is thus a striking contrast between a slow, massively parallel brain and a fast, highly repetitive computer. But there are other differences in the ways the two function.

2. WE HAVE POOR MEMORIES

It is possible for some people to learn by heart long literary or religious texts, like Homer's *Iliad* and *Odyssey* or the Bible. But computers typically can do much better: the text of the *Encyclopaedia Britannica* fits easily on a CD-ROM or on a modern hard disk. Of course one should beware of concluding too rapidly that computers are superior to us. Learning long texts by heart is after all not the main use of our memory, and it is hard to quantify what our memory is really good at. We want nevertheless to argue that human memory is not very good for doing mathematics. Humans (like computers) have several types of memory, but it seems sufficient for our purposes to distinguish between long-term and short-term memory. It takes time to enter things in our long-term memory (this apparently requires protein synthesis): you won't remember a long string of random words or digits after you read them once. Short-term memory is what lets you remember a list of items that has just been presented to you, and it is limited to about *seven* items. This means that it may be a difficult task to read the digits of a telephone number and then dial without looking again at the directory. Dialing telephone numbers efficiently did not have significant survival value until recently, otherwise natural selection might have made us more adept at this task.

Things being as they are, mathematicians put a lot of facts in their long-term memory through long days of study. After which, they remember the definition of the cross-ratio, the fact that it is preserved by projective transformations, and many other things. As for short-term memory, it is supplemented by the use of the blackboard, or a sheet of paper, or a computer screen. These serve as external memories, readily consulted by looking at them. Long-term memory can also be complemented by books and other visual media. This brings us to the next point.

3. The Human Brain Has Well-Developed Visual and Linguistic Abilities

Our visual system has evolved over many millions of years to a remarkably efficient instrument. In a fraction of a second we spot and recognize an animal or object hidden in a complex background. Clearly, this ability has been important for the survival of our ancestors. But now we can use it to stare at geometrical figures, diagrams, formulas, and mathematical text. If we had to build a computer with mathematical ability, we would probably not start with a very complex visual system. But we humans have this wonderful instrument at our disposal, and we naturally use it in doing mathematics. In other words, the way we do mathematics is strongly influenced by the use of our complex and efficient visual system.

The human ability to communicate complicated abstract information through language is recent from an evolutionary point of view (dating from perhaps 50,000 years ago). Its survival value is evident and explains the huge number of humans currently crowding our planet. The use of a human natural language is central to human mathematics.[4] The language may be ancient Greek, modern English, or something else, spoken or written. But basically, all that we call mathematics uses a natural language, even if mathematicians insist that mathematical texts could in principle be written in a formal language. In practice, formal languages are not used and cannot be used. Our natural languages are powerful and versatile indeed. The fact that we have to rely on them is a peculiarity of human mathematics and, in fact, a shortcoming, as it makes impossible any mechanical checking of the correctness of mathematical texts. This is connected with the last point that we want to discuss in this chapter.

4. HUMAN THINKING LACKS FORMAL PRECISION

One task that a computer performs with the utmost ease is comparing two long files and deciding if they are identical or not. The files may contain the text of a novel in Irish or Icelandic, and in a fraction of a second the computer will tell you if a word has been spelled differently in one of the two files. For a human, the task would be long and arduous, and would depend on irrelevant details like whether you understand Irish or Icelandic. If the text of a novel is replaced by that of the *Encyclopaedia Britannica* or a full current set of U.S. telephone directories, the task now borders on the impossible for a human, while remaining easy for a computer.

The above example shows how a logically easy task, if it is very long and has to be performed without any error,[5] is difficult for humans and easy for a computer. This is certainly a shortcoming for us in doing mathematics. Of course when it comes to recognizing a spade or a cat when we see one, we do much better than current computers. And our superiority is crushing in the domain of mathematical creativity. But you will agree, I think, that our way of attacking mathematical problems is somewhat idiosyncratic, and if a colleague from outer space came to visit us, she might be somewhat perplexed by how we proceed.[6]

❖ 10 ❖

Mathematical Texts

Wᴇ ꜱᴘᴇᴀᴋ ᴏꜰ mathematical reality as we speak of physical reality. They are different but both quite real. Mathematical reality is of logical nature, while physical reality is tied to the universe in which we live and which we perceive through our senses. This is not to say that we can readily define mathematical or physical reality, but we can relate to them by making mathematical proofs or physical experiments. We can also make guesses, and reality can confirm or falsify these guesses. The relations of the human mind with mathematical or physical reality are complex. The comparison that we have made in Chapter 9 of the computer and the brain revealed some of the subtleties that underlie the application of human thought to mathematics. We want to take now a different point of view and examine the end product of mathematical activity: the mathematical text.

Before I discuss written mathematical texts, let me mention an important variation: the oral presentation, which may be a lecture, seminar, colloquium, or the like. In an oral presentation the mathematician stands and writes on the blackboard while speaking for about an hour. These days the blackboard is often replaced or supplemented by an overhead projector, used to display on a screen some transparencies prepared in advance. Note that there are important differences between mathematics and other disciplines with respect to oral presentation. A philosopher might sit at a table and read a carefully prepared text. A physicist might use a computer to project on the screen some colorful pictures and text, possibly animated. But mathematicians like the traditional use of chalk and blackboard (or innocent variations like a whiteboard). This setup has the advantage of limiting the amount of information received by the audience per unit time. The fact is that there is only so much information that can be transmitted in one lecture. Flashing complicated formulas at great speed on a screen or speaking for two hours instead of one is rather useless and makes everyone dizzy. (This is again a difference between the computer and the brain: a properly plugged in and programmed computer can "think" for days in a row without need for a nap or a cup of coffee.)

Written mathematical texts may be books or articles (also called papers) of various lengths. The articles are published in specialized journals and/or, these days, posted on the Internet. A written mathematical article is the basic end product of human mathematical activity. It can permanently be consulted and checked for correctness. A new mathematical idea acquires legitimacy only when you have written it down and published it.

For the purposes of the present discussion, we can see a mathematical text as composed of three kinds of components: figures, sentences, and formulas.

FIGURES

In the geometry of Euclid an important role is played by figures,[1] and constructions made on the figures (such as *draw the perpendicular to AB through the point C* . . .). Figures put the human visual system to good use and constitute a valuable external memory as soon as the geometric situation considered is a bit complicated.[2] Reasoning on figures is formidably effective, and this explains how geometry was historically the first branch of mathematics in which really profound and difficult results were obtained.

Nevertheless, you will often find no figure at all in a modern mathematical paper, even if the subject is a question of geometry. The main reason for this disaffection is that one can make mistakes by relying too much on a specific figure when trying to prove a general result. Reasoning on a figure is thus discouraged as nonrigorous. Figures remain useful, however, to fix attention and as external memories, and they are used a lot in oral presentations.

Suppose that a seminar speaker utters the sentence, "We consider a geodesic arc joining the points A and B of the Riemann manifold M." At the same time he or she would draw this on the blackboard:

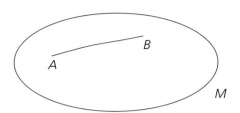

In a mathematical paper the sentence would be written without a picture, but many readers would have a picture in their mind.

The absence of figures, therefore, does not mean that geometric intuition is shunned. On the contrary, *geometrization* is welcomed by mathematicians: this consists in giving a geometric interpretation of mathematical objects (in algebra or number theory) that are a priori nongeometric.

While visual intuition is important, we must admit that it has no logical necessity in mathematics. Some people do not use it, and the French mathematician Laurent Schwartz (1915–2002) claimed to have so little ability in this respect that it was very hard for him to use a road map. This raises the fascinating question of the variety of internal representations of mathematical objects by various mathematicians. Too little is known on this topic, and I shall not dwell on it here.

Sentences

We stated above a typical mathematical sentence, "We consider a geodesic arc joining the points A and B of the Riemann manifold M." This sentence is in English, with some symbols (A, B, M) and some jargon (geodesic arc, Riemann manifold). It would be easy to translate the sentence into French, German, or other natural languages. But, as said earlier, some natural language is necessary to do mathematics in practice, even if not in principle. One can do mathematics without figures and without formulas, but one needs to use a natural language.

Language plays a central (although not exclusive) role in human thinking. But language is a rather diverse thing, and its use in mathematics is very different from its use in poetry. How is that? Instead of the mathematical sentence discussed above, let me write, "Let N be a Riemann manifold, and take a geodesic arc AB in N." We have given a new name N to the Riemann manifold and otherwise said the same thing with other words. In poetry, if you change names and use other words you are NOT saying the same thing. Think of Edgar Allan Poe's poem "The Raven." Replace the name Lenore (which rhymes with *nevermore*) by some other name, like Madison, and modify some words and grammatical forms without altering the meaning. Soon enough you will have changed a powerful poem into inane rambling. Obviously reading poetry and reading mathematics are rather different activities of the brain. For poetry the regularities and irregularities in the form are an essential part of the message.[3] For mathematics, by contrast, the form is of limited importance. If you are bilingual and discuss an idea with a colleague of the same persuasion, you may remember afterwards which was the mathematical topic of your conversation but not which language you used.

FORMULAS

Mathematical texts are usually interspersed with formulas, like

$$\frac{U-A}{M-A} : \frac{U-B}{M-B} = \frac{M-A}{V-A} : \frac{M-B}{V-B} , \qquad (*)$$

which we met in chapter 4. There is no fundamental difference between a formula and a sentence. In fact you can pronounce the formula that we have labeled (*) as, "U minus A over M minus A divided by . . . is equal to. . . ."[4] To most mathematicians the formula is highly preferable to the sentence. For this I see two essential reasons. First, we can apply our visual skill to the formula, as we could to a geometric figure, and treat it as visual memory. Second, there are rules by which we can mechanically obtain one formula from another with relatively little effort and little risk of mistakes. In the case of formula (*) we used the extra information that M is the midpoint of AB, which can be expressed as

$$(M-A) : (M-B) = -1. \qquad (**)$$

Inserting (**) in (*) one finds

$$(U - A)(V - A) = (U - B)(V - B).$$

(To a trained mathematician this is immediately *visible*.) Then easy manipulations give

$$\frac{U + V}{2} = \frac{A + B}{2} = M,$$

and that was our proof of the butterfly theorem.

Systematic and easy manipulation of formulas is an essential addition of modern mathematics over the intellectual tools available to the ancient Greeks. Formulas, as was the case for figures, are often in the mind of mathematicians even when not explicitly written out. Note also that formulas do not always relate numbers (complex numbers in the case of (*)). The formula $A \subset B$ says that the set A is contained in the set B, and there are formulas to express any other logical relation. Whatever its meaning, a written formula has in principle the value of an external memory and of an object that can be usefully manipulated according to well-defined rules.

We have said and repeated that mathematics can *in principle* be presented without the use of a natural language. Such a presentation of mathematics would be "formulas only" and could be checked mechanically for correctness. In fact, some mathematicians (in particular, beginners) like to write formulas rather than sentences because they think it is more "rigorous." But this practice rapidly produces an incomprehensible mess. Efficient transmission of mathematics to humans depends on lucky choices for what to express with formulas and what to express with words (which refer to unwritten formulas). Making these lucky choices is a different ability from pure technical power. It is an art, and some mathematicians are very much better at it than others.

❖ 11 ❖

Honors

In EARLIER CHAPTERS we have been asking about the nature of human mathematics. Let us now pause and ask why humans get involved in mathematical research or scientific research in general. (Mathematics is more of a pure intellectual game and of a solitary enterprise than research in other fields, but these are differences of degree rather than of nature.) Of course people go into research for the challenge, the interest of the subject, for the money, for fame. . . . Looking more deeply into these motivations may not teach us much about the structure of mathematical reality, but it can tell us something about human nature and society. In this chapter we shall be modest and ask only about the more visible part of the emotional relation between scientists and science.

My colleague the theoretical physicist Louis Michel liked to say that people choose an academic career from lack of imagination. How is that? Suppose you do well in school. Then a natural idea is to try to go to college. Then if you have no particular urge to jump into "real life" (whatever that is), you will proceed to a PhD. And when you are through with that, academic life will probably be real life for you. It will be hard to imagine something else. Through lack of imagination you will try to stay in academia and "do research," among other things. Now, good research requires finding new solutions to new problems, and this is a definition of imagination or of intelligence. But good research also requires a large amount of routine work, often delicate and complex, that has to be done with care, with precision, and without undue imagination. Lack of imagination is thus needed for research, even if some imagination is also needed to do good research.

At the root of science and scientific research is the urge, the compulsion, to understand the nature of things. We shall want later to look into this compulsive behavior and try to make sense of it. But science has other motivations (like lack of imagination as we just saw). And an important motivational aspect of human science is its reward system, which we now want to discuss.

Human beings, like other animals, have an inborn system of instincts or drives (later modified by experience) that govern their unconscious and conscious activities with respect to breathing, food, sex, and so on. These drives have in part been analyzed at the physiological level. They are often associated with what we consider pleasant or unpleasant stimuli. For instance, it appears that a substance called *histamine*, released from cells by one or another form of injury, plays a role in the initiation of sensory impulses evoking pain or itch. One should, however, beware of an overly simplistic and mechanical explanation of drives as responses to remembered pleasant or unpleasant stimuli. If the antelope runs away from the lion, it is not because it remembers being eaten by the lion as an unpleasant experience! Keeping away from dangerous beasts is hard wired, at least in part, in the brains of animals and humans. And this is presumably why many of us stay away from snakes, spiders, and angry dogs. Besides that, man is a social animal, and we conform, more or less, to the rules of the group in which we live.

Among the incentives to behave one way or the other is what we were taught by our parents. Instead of our parents it may be some mother or father image or indeed God who tells us what to do. God is the father image par excellence. The father image typically is benevolent, as long as we comply with his wishes, otherwise his wrath is terrible. The mother or father image will try to enforce her or his dominant role, which we just have to accept. Rejecting it is a sin, a crime, and a very dangerous thing. Remember Giordano Bruno[1] being burned at the stake in Rome for his philosophical opposition to the doctrines of his holy mother, the Catholic Church. And remember the countless millions of victims of fatherly totalitarian regimes, right wing or left wing, religious or antireligious.

Free discussion as it occurs in science is not the common right of women and men. Free thinking about philosophical issues and questioning religion or social structures has always been the exception rather than the rule. The rule is to respect the surrounding power structure and ideology. Power and ideology change with place and time, and we can help change them, but they are always present. This presence may be upsetting, but mostly we take it for granted. If we live in a democratic country, with a reasonably liberal regime, we may find it easy to accept the power structure and ideology of our community. But moving from one place to another, we see that the rules are different: where the president of the United States feels obliged to make frequent references to God, the French president is not allowed to do so.

At the risk of belaboring the obvious, let me repeat the argument. Power structure and pressure to conform may take the most detestable forms. But detestable or not, they are part of the fabric of human society. Power structure and pressure to conform may be enforced by brutal methods, but they are also grounded in basic human psychology and our acceptance of dominant mother and father images. As we can expect, this also applies to science. There is more freedom of discussion in science than in other domains, but there is a power structure and there is a pressure to conform. The power structure is expressed in hiring and salaries, the pressure to conform in acceptance of papers for publication in scientific journals. Overall the system works in a reasonably efficient and satisfactory manner. It can certainly be improved, but I am not advocating its destruction. One must also mention here the admiration that some scientists earn by the results they have obtained and the honors that, in some cases, some father image bestows upon them. Scientific honors are often a matter of deep emotion and of pronounced irrationality. Examples abound of truly great scientists whose life is made miserable because they did not get some desired honor, even when a rational analysis would reveal this honor to be more of a nuisance than a blessing.

What I call an honor is being elected to an academy, winning some prize or medal, being chosen to give some special lecture, or simply being offered a prestigious professional position. Honors provide rewards in various proportions: ego satisfaction, money, political power, professional assistance, and obligations that may be time-consuming. There are thus material and psychological, rational and irrational aspects to receiving honors. But there is also politics above the level of the individual being honored. For example, fund raising is facilitated when a university can boast several Nobel Prize winners on its faculty: they play the same role as (in the United States) a good football team.

There is no Nobel Prize in mathematics, and I can remember a time, in the 1960s and 70s, when mathematicians were quite happy about that. Mathematical quality was not measured in millions of dollars, and mathematicians were not rated with football players. There is in mathematics a fairly prestigious Fields Medal, but it is given only to mathematicians below the age of forty and has negligible financial value (unlike the physics Nobel Prize, worth about $1 million). As I remember it, the attribution of the Fields Medal to René Thom and Alexander Grothendieck was not a very big deal. (At one point René Thom was mildly annoyed because his wife, Suzanne, had misplaced his Fields Medal, and it could not be found).

But things have changed. The Fields Medal is now commonly referred to as the Nobel Prize of mathematics. And several other prizes are now given to mathematicians, worth about $1 million and which claim to be the Nobel Prize of mathematics. Million-dollar prizes have also been promised for the solution of some outstanding mathematical problems. Many mathematicians are happy to mention a "million-dollar problem" when the occasion arises. Others feel that putting a price tag of $1 million on the Riemann Hypothesis is sort of tasteless. Certainly if the unit of mathematical excellence is one million dollars, this is far below the unit of excellence for golf, tennis, or car racing.[2] But let's not lose too much time on this million-dollar issue.

My opinion is that honors currently play an excessive role in mathematics. (Perhaps this will change again.) The reason for the big role is probably the difficulty of evaluating contemporary work, which is often very technical and hard to understand. Instead of explaining the theorem proved by X, it is easier to say that X has received the Alpha Prize. But one must admit that intellectually this is less interesting, whether or not the Alpha Prize has been fairly attributed. In fact, the Alpha Prize is not attributed by God Almighty, but by a committee that has received nominations, solicited and read reports, and does not always make a good choice. The selection of a scientist for some honor may officially be restricted by citizenship, age, and other such categories, but it should not depend on ethnicity, gender, political views, and so on. It is, however, a fact that the latter often also play a decisive role.

Let us look at one example of evaluation. A good math department (say, Princeton University) wants to hire a new faculty member. The choice here is made by competent mathematicians, able to find and rate the talented young people. The department wants to hire colleagues of high mathematical ability and will probably make a good choice. Consider by contrast a committee that does not feel very competent but wants to defend its own respectability. The committee may play it safe, look for a candidate who has already received various honors, and give him or her one more. Or it may follow rumors that a candidate has almost proved some important result, and make a bad choice.

I have tried to show that evaluation of scientists, and honors, may work out well or not so well, but they are a necessary part of human science. They may be a nuisance, but they are a serious problem that cannot be dismissed or ignored. Serious problems, however, should often not be taken too seriously. And relief sometimes comes from unexpected quarters, as the following story shows.

I had been attending a solemn meeting of the Académie des Sciences de Paris, under the cupola of the Institut de France. The audience was distinguished, and many of those present wore the green costume of academician, with cocked hat in hand and sword at the side. There were elegant speeches about aspects of the life of the Académie. Then other addresses were delivered about such weighty questions as the future of mankind and our planet, the responsibility of the scientist, and so on. The meeting came to an end after a couple of hours, and as there were other things that I still wanted to do, I prepared for a discrete and speedy exit. I was the first outside. But, as I rushed, I sensed that something was wrong: I was between two rows of brilliantly adorned soldiers with drawn sabers. This was *la garde républicaine*, and they were supposed to salute the green costumes of academicians at the beat of the drums. (No trumpets, that's for the day of judgement.) My timing and my attire were wrong. I should have gone out the side rather than between the fierce-looking soldiers. I tried to look inconspicuous, but how do you look inconspicuous while zigzagging between two rows of *gardes républicains* with drawn sabers? I kept glancing left and right to keep my bearings. And then I caught sight of the chief of the *gardes*: a giant. He was formidable, noble looking, and had an impassible face from which I could not detach my gaze. He looked at me too, unsmiling. Then I saw him close one eye, and slowly open it again: an unmistakable wink that saved my day.

❖ 12 ❖

Infinity: The Smoke Screen of the Gods

LET US RETURN NOW to the essential activity of mathematicians: proving theorems by applying rules of logic to other theorems and to the basic axioms. All of mathematics can be developed from set theory, so that all we need are the axioms of set theory. What are they? Most of the mathematics done today is based on a system of axioms called ZFC (or Zermelo-Fraenkel-Choice, for Ernst Zermelo,[1] Adolf Fraenkel,[2] and Axiom of Choice). In fact, mathematicians rarely use the ZFC axioms as such: they invoke well-known theorems that can be derived from ZFC. So, if you want to prove that *there are infinitely many primes*, you will not usually try to derive this fact from ZFC. You will instead rely on the notion that someone has already established a connection between integers and set theory, and derived for you a certain number of well-known facts about integers (see below).

You can look up the ZFC axioms in various places. Using the *Encyclopedic Dictionary of Mathematics*[3] I found a list of ten axioms formulated in formal language. Axiom 5 is

$$\exists \, x \, \forall \, y \, (\neg \, y \in x).$$

Remember that you have to know "rules of logic" that allow you to manipulate the symbols that constitute axioms and theorems. There is also an intuitive meaning to the formal expressions that you write. You can in principle do mathematics without the intuitive meaning, but the intuitive meaning is usually considered essential by human mathematicians. As to Axiom 5, it says that there exists a set x such that for all y it is false that y belongs to x. In other words there is a set x that contains no element. The set x is called the empty set and is usually denoted by $\emptyset$. Axiom 5 thus says that *there exists an empty set* $\emptyset$. Once you have the set $\emptyset$, you can also consider the (different) set $\{\emptyset\}$, which has just one element (namely, the empty set $\emptyset$), and the set $\{\{\emptyset\}\}$ with one element $\{\emptyset\}$. You may also consider the set $\{\emptyset, \{\emptyset\}\}$ containing the two elements $\emptyset$, $\{\emptyset\}$, the set $\{\emptyset, \{\emptyset\}, \{\{\emptyset\}\}\}$, containing three elements, and so on. If this makes you slightly dizzy, don't worry: it is a perfectly normal reaction. But notice in passing that we have discovered a way of introducing the natural integers 0, 1, 2, 3, . . . by associating them with the sets $\varnothing$, $\{\varnothing\}$, $\{\varnothing, \{\varnothing\}\}$, $\{\varnothing, \{\varnothing\}, \{\{\varnothing\}\}\}$, ...

This is not the place for a detailed discussion of the ZFC axioms. But let me mention Axiom 6, which, in ordinary language, says that there is a set with infinitely many elements or as the mathematicians like to say, *there exists an infinite set*. Why should one insist on such an axiom, when it is clear that the natural integers 0, 1, 2, 3, . . . form an infinite set? The point is that the axioms come first and the integers later, and what appears intuitively clear is a minor consideration when you want to provide solid foundations to mathematics. The grandiose construction of abstract set theory initiated by Georg Cantor at the end of the ninteenth century was plagued with paradoxes that led to a crisis of the foundations of mathematics.[4] We must be thankful to the logicians, who, at the beginning of the twentieth century, did a good job of giving us an axiomatic basis for set theory.

But I forgot to say what an infinite set is. An informal definition is that a set is infinite if it contains as many elements as a strictly smaller subset. For instance, the set $\{0, 1, 2, 3, . . .\}$ formed by the natural integers contains as many elements as the subset $\{0, 2, 4, . . .\}$ formed by the even natural integers. (To see this, associate the integer n and its double $2n$.) But the set of the even natural integers is strictly smaller than the set of all natural integers because it is missing 1, 3, 5, ... In conclusion, the set $\{0, 1, 2, 3, . . .\}$ of the natural integers is infinite. And now that we have defined what *infinite* means, it also makes sense to say that there are infinitely many primes.

If two primes differ by 2, they are called twin primes (3 and 5 are twin primes, and similarly 5 and 7, 11 and 13, 17 and 19, . . .). It is believed, but not proved, that there are infinitely many pairs of twin primes. So, while the ZFC axioms give a satisfactory foundation to our mathematics, that does not mean that all apparently reasonable questions are easily answered. In fact, Gödel's incompleteness theorem says that there is no way to give systematically an answer to all mathematical questions.

But let us forget Gödel for the moment and try to solve the problem of twin primes by brute force. We compute all the primes less than some value N (this can be done quite explicitly), then we pick out the pairs of twins (this can also be done quite explicitly), and then we are stuck: we would have to do the calculation for arbitrarily large N, and this would take an infinite time. The solution of the problem of twin primes is *hidden at infinity* in arbitrarily large values of N.

How is it then that we know that there are infinitely many primes? The answer is that we do not try to compute all primes but use instead a clever mathematical argument. The argument was known to Euclid: write $n! = 1.2 \ldots n$ for the product of all consecutive integers from 1 to n. Clearly, $n!$ is an exact multiple of each integer k from 2 to n. Or, as one says, k *divides* $n!$ (i.e., the remainder of the division of $n!$ by k is 0). But k does not divide $n! + 1$ (because here the remainder of the division is 1). Therefore, any number k (greater than 1) that divides $n! + 1$ must be greater than n. In particular any prime factor of $n! + 1$ is greater than n. For arbitrarily large n we can thus find a prime k larger than n: there are arbitrarily large primes. Note that in the argument just given we did not go back to the ZFC axioms. We used instead *informally* a certain number of well-known concepts and facts about integers (like the fact that an integer can be written as a product of primes in a unique way, apart from the order of the factors). This is the usual way mathematicians proceed. But in principle it is possible to go back all the way to ZFC, to relate integers to set theory, and proceed with absolute rigor.

The beauty of mathematics is that clever arguments give answers to problems for which brute force is hopeless, but there is no guarantee that a clever argument always exists! We just saw a clever argument to prove that there are infinitely many primes, but we don't know any argument to prove that there are infinitely many pairs of twin primes.

Let us now try another idea. Starting from your favorite axioms (say, ZFC), you can write systematically and mechanically a list of all correct proofs. You can thus also write a list of all statements that have a proof from your axioms, checking at each step if you have obtained a proof of your favorite statement (like *there are infinitely many pairs of twin primes*). You will use formal language (like that in Axiom 5) for your list of statements that have a proof, and you will have an *algorithm* to produce the list systematically and mechanically. (An algorithm leads you through a sequence of steps, telling you exactly what to do at each step. You can implement an algorithm with a suitably pro-grammed computer.) Note that in the list of statements pro-duced by your algorithm, statements may be repeated, and some short statements may appear quite late and unexpectedly.

So, there is an algorithm that produces a list of statements provable from your axioms. Here is, however, a remarkable (and nonobvious) fact: *there is no algorithm that produces a list of all statements that cannot be proved from the axioms.*[5] In the jargon of mathematical logic, one says that the set of provable statements is *recursively enumerable*, while the set of statements that have no proof is not recursively enumerable. Note now that the set of statements that have a provable negation is again recursively enumerable and therefore cannot coincide with the set of state-ments that cannot be proved. This is Gödel's incompleteness theorem: *if a theory is consistent (i.e., one cannot prove a statement and its negation), then there are statements that can neither be proved nor disproved.*[6]

From what we have just seen, there is a relation between algo-rithms and Gödel's incompleteness theorem: some collections of statements (or of natural integers) can be produced by an algo-rithm and some cannot. This has to do with the fact that there are infinitely many statements (or integers). When dealing with infinite sets, there are limitations as to what tasks can be *effec-tively* performed.

What are the tasks that can be effectively performed? Different answers have been given by Gödel, Church,[7] Turing, and these answers turn out, fortunately, to be equivalent. In brief, a task can be effectively performed if there is a computer that can per-form it. This computer is a finite automaton (Turing showed that it could be quite simple) with unlimited memory and unlimited time at its disposal.

Before we leave Gödel, let me mention a consequence of his work that is relevant to the actual practice of mathematics: short statements can have arbitrarily long proofs. What does this mean? Specifically that, as the length L varies, the maximum, for a given L, of

length of the shortest proof of a provable statement of length L

is not an effectively computable function of the length L.[8] Since the functions you know (polynomials, exponentials, exponentials of exponentials, etc.) are effectively computable, this means that the length of the proof grows faster with the length of the statement L than anything you can lay your hands on.

To repeat: some short statements have a proof that is very long and thus is hard to find, and you don't know if there is a proof at all until you find one. Such proofs are therefore highly valued by mathematicians.

You may at this point wonder what kind of game the mathematicians are up to, introducing counterintuitive notions like sets that cannot be constructed by an algorithm or functions that are not effectively calculable. Is this really necessary? That depends on what you want to do. Thousands of years ago, it became important for our ancestors to count heads of cattle, slaves, or bushels of grain. And bartering one kind of commodity for another led to questions of elementary arithmetic that were, at the time, not very easy to solve. But noncalculable functions did not pop up then. Some of our ancestors, however, went beyond counting sheep and exchanging them for jugs of oil or wine: they started thinking about numbers in general, about all possible triangles or other geometric shapes. When that happened at some time in antiquity, mathematics was born. To discuss general properties of numbers or triangles, one cannot use the brute force methods of looking at each of them one after the other: there are too many. By discussing all the elements of the infinite set of numbers or geometric figures, man is stepping into the domain of the Gods (as Plato would see it). And in the domain of the Gods, many wonderful mathematical facts have been found: hidden properties of integers, unexpected theorems of geometry, and so on. But not all mysteries have been unveiled: there remain questions that will perhaps be solved in the future. Or perhaps they cannot be solved, and because of Gödel's incompleteness theorem we shall never know. Mathematicians want to talk about properties of all elements of infinite sets. But in an infinite set, things can be hidden far away. And Plato might have been pleased to see that, while the Gods have allowed us in their domain, they have found a way to keep some of their mysteries inaccessible to us.

❖ 13 ❖

Foundations

THE MATHEMATICS of antiquity dealt with objects that appeared natural, namely, geometric figures and numbers, and had a natural method, namely, logical deduction from a few accepted axioms and definitions. The principle of the axiomatic method has remained basically unchanged even today, but the wealth of objects considered has increased enormously, as language and techniques have become more diverse.

A modern description of mathematics, à la Bourbaki, would put weight on *structures*: simple ones like the structure of *groups*,[1] or more complex ones like the structure of *algebraic varieties* mentioned earlier. The simple structures would be studied under the heading of set theory, algebra, topology, and so on, while more complex structures would come under the heading of algebraic geometry or smooth dynamics, for example. (Groups are part of algebra, while algebraic varieties are part of algebraic geometry.) This classification of mathematical topics is certainly convenient but a bit bureaucratic, and its naturalness remains to be assessed. In fact, interesting mathematics develops along lines that may or may not respect the structural views of the Bourbakists.

The diversification of topics discussed in modern mathematics has been balanced by unifying tendencies. One unifying factor is the unexpected appearance of connections between apparently unrelated questions. For instance, a topic known as the *theory of functions of a complex variable*[2] has turned out to be an essential tool in a completely different area of mathematics: arithmetic (the study of integers). In fact, the most famous open question in mathematics today is the Riemann Hypothesis, a conjecture about properties of a particular function of a complex variable that would have important consequences for our knowledge of prime numbers.[3]

Another unifying factor of mathematics is that all of it can be based on an axiomatic treatment of set theory, as we have seen in chapter 12. Here again the basic axioms (say, ZFC) have an intuitive meaning that make them acceptable to modern mathematicians, just as Euclid's axioms were acceptable to the Greeks.

But while mathematics of antiquity was appealingly natural, one cannot say the same of our current mathematics. While the ancient mathematicians were looking for truth, we seem to be looking for consequences of axioms that could be replaced by other axioms with different consequences. While the ancients played with straight lines, circles, and integers, we have introduced a plethora of esoteric structures. And if one looks at the technical journals where modern mathematicians record their studies, one may well wonder what they are up to: why choose this problem? why make those assumptions? what is the point of it all?

What we would like to understand is how natural our current mathematics is. I see two aspects to this question: first, the problem of arbitrariness of the foundations (why ZFC?), then the problem of the arbitrariness of the questions studied (why Fermat's last theorem?).

Let me start in this chapter with the problem of foundations and accept the view that all of mathematics is based on set theory.[4] This is a widely accepted view in today's practice, even if mathematicians of a later period will perhaps see things quite differently. But what about the specific choice of the ZFC axioms?

It is instructive to discuss the case of the Axiom of Choice (the C of ZFC). What the Axiom of Choice says is not essential for the present discussion and is therefore left to a note.[5] A strange consequence of the Axiom of Choice, the Banach-Tarski paradox, is left for another note.[6] Let us call ZF the axioms of Zermelo-Fraenkel without the Axiom of Choice. It was proved by Gödel that if ZF is consistent (i.e., no contradiction arises from these axioms), then ZFC is also consistent.[7] Consistency is thus not an issue in using the Axiom of Choice, but there are other considerations. Some mathematicians frankly dislike the Axiom of Choice, and others make a mental note whether it is used in a given theory. But at this time most mathematicians consider that they obtain richer, more interesting mathematics with this axiom than without.[8] This makes ZFC currently the standard basis for mathematics.

Does this mean that the axiomatic basis of mathematics will no longer be changed in the future? I personally think that there will be changes, but they will happen slowly.

Among the problems that mathematicians have tackled in the last hundred years, some have been brought to a satisfactory solution, like the proof of Fermat's last theorem or the classification of finite simple groups.[9] These successes were achieved at the cost of very long proofs. (This is not too astonishing in view of Gödel's results on the length of proofs mentioned in the last chapter.) Some problems have been proved to be logically undecidable: this is the case of Hilbert's tenth problem, on Diophantine equations.[10] Finally, some problems are still wide open, like the Riemann Hypothesis.

The Riemann Hypothesis (RH) is a technical conjecture that, for several reasons, has captured the attention or even passion of mathematicians. The conjecture is about a certain function called the *Riemann zeta function*, and it is relatively easy to formulate precisely. We give the standard formulation of RH in a note,[11] but this does not begin to show why RH is interesting. The first reason why one would like to know if RH is true was given by Riemann himself: RH entails detailed results on prime numbers, which appear inaccessible otherwise but are believed to be true. A second reason for our interest in RH is that it appears extraordinarily difficult to prove. A last and most important reason is that RH is tied to deep structural questions. In particular the Weil conjectures (which were mentioned earlier and which have been proved by Grothendieck and Deligne) contain an idea related to RH but in an apparently unrelated setting. While a technical discussion is beyond the scope of this chapter, we will now have a look at some logical issues involved in RH, which will give us an idea of the way of thinking of logicians.

Technically, RH says that the Riemann zeta function is never zero in a certain "forbidden region" of the complex plane where it is defined. In particular, if RH is *false*, one can prove that it is false by exhibiting a *zero* in the forbidden region. (This can be done by a numerical computation.) Suppose now that RH is *undecidable*. Because of undecidability one cannot exhibit a zero in the forbidden region. (Indeed, this would show that RH is false and thus not undecidable.) But if one cannot exhibit a zero, this means that there *is* no zero in the forbidden region; that is,

RH is true! More precisely, if RH is undecidable on the basis of ZFC *and* ZFC is consistent (does not lead to contradictions), *then* RH is true.

We have said that RH is testable by numerical computations. Indeed, a lot of work has been done already, leading to considerable numerical evidence in favor of RH. But while a numerical computation could firmly demonstrate that RH is false, the sort of evidence we have cannot prove that RH is true. One may view the numerical computations just discussed as a lot of effort to *disprove* RH, unsuccessful so far. There has also been a lot of (nonnumerical) effort to *prove* RH, also unsuccessful. It seems safe to say that no clever hypergenius will come up tomorrow with a short proof, or disproof, or RH. If there is a proof, or disproof, it is likely to be long and difficult. A sobering thought: RH might have a proof, but it may be so long that the physical limitations of the universe in which we live would prevent any implementation of the proof. (It would require too much paper, or too much computer time. . . .)

But let us come back to the possibility that RH is undecidable; that is, no proof or disproof exists. Apparently this possibility is the end of the road; there is nothing more we can do. But actually there is: one can try to *prove* that RH is undecidable. It is not inconceivable that this can be done, and the logician Saharon Shelah has in fact put forward what he prudently calls a *dream*: *prove that the Riemann Hypothesis is unprovable in PA but provable in some higher theory.*[12] The initials PA here stand for Peano Arithmetic (see note 4), a weaker system of axioms than ZFC. Shelah's idea is to use the techniques of mathematical logics to prove the undecidability of RH over PA. It then follows that RH is true if PA is consistent.

Mathematical logicians, looking at axiom systems from outside (i.e., doing metamathematics), can thus achieve an understanding that escapes the standard mathematician working inside an axiom system like ZFC or PA. It is, however, a sociological fact that, at this time, most mathematicians view metamathematics with a certain lack of enthusiasm. They pay due respect to Gödel and his incompleteness theorem, hail the proof that Hilbert's tenth problem is unsolvable, but prefer to do the sort of *real mathematics* for which they have developed a refined set of techniques, intuition, and taste.

Things are changing, however. Today we might almost say that mathematics consists of studying the consequences of ZFC. But one can doubt that this will still be the case one century from now. And whether we like it or not, it seems that mathematical logic (metamathematics) will play an important role in the future of our mathematics.

❖ 14 ❖

Structures and Concept Creation

W E HAVE SEEN in chapter 3 that the idea of mathematical structure is present in Klein's Erlangen program. In a different form, structures dominate Bourbaki's *Eléments de Mathématique*. One might say that structures permeate modern mathematics, sometimes explicitly, sometimes not. Yet the jump from the axioms of set theory to the definition of various structures (like groups, geometries, etc.) may seem a bit contrived: we would like to understand if the choices that are made are inevitable and natural.

But before discussing the origin of structures, let me clarify a point of terminology concerning axioms. When we introduce the concept of a group, we do this by imposing certain properties that should hold[1]: these properties are called *axioms*. The axioms defining a group are, however, of a somewhat different nature from the ZFC axioms of set theory. Basically, whenever we do mathematics, we accept ZFC: a current mathematical paper systematically uses well-known consequences of ZFC (and normally does not mention ZFC). The axioms of a group by contrast are used only when appropriate. Suppose we have, for the purposes of the problem on which we are working, introduced a product $a \cdot b$ between elements a, b of some set G. If this product satisfies the properties appropriate for a group (i.e., associativity, existence of unit element, and existence of inverses), we say that the set G with the product $\cdot$ is a group. Or it may happen that some axiom (e.g., associativity: $a \cdot (b \cdot c) = (a \cdot b) \cdot c$) is not satisfied, and then G is not a group.

Euclid's axioms played for the Greeks a role similar to that of ZFC for us. But nowadays Euclidean geometry is approached differently. Starting from ZFC, one defines real numbers, then the Euclidean plane (or the Euclidean three-space). And then one can verify that points, lines, and so on verify the axioms introduced by Euclid or the reworking of these axioms by Hilbert.[2] In this approach, Euclidean geometry is thus a derived concept.

Let us go back to structures. To discuss their significance in mathematics, we have to remember the dual nature of the subject. On the one hand, mathematics is a logical construct that can be identified with the output of a Turing machine working forever at listing all consequences of the ZFC axioms. This is the mechanical, completely nonhuman aspect of our subject. On the other hand, mathematics is a human activity among many others. Suppose that you are a rock climber. Climbing involves you, a human, and the rock on which you are progressing, which is mineral and completely nonhuman. The ledge on which you hoist yourself was not waiting for you: it is a result of erosion acting on compacted sediments originally deposited on a sea bottom many million years ago. Yet this ledge, on which you now precariously stand, and the handhold, fortunately excellent, that you just found have a lot of meaning for your human nature: your life depends on them.

It thus stands to reason that mathematical structures have a dual origin: in part human, in part purely logical. Human mathematics requires short formulations (because of our poor memory, etc.). But mathematical logic dictates that theorems with a short formulation may have very long proofs, as shown by Gödel.[3] Clearly you don't want to go through the same long proof again and again. You will try instead to use repeatedly the short theorem that you have obtained. And an important tool to obtain short formulations is to give short names to mathematical objects that occur repeatedly. These short names describe new concepts. So we see how concept creation arises in the practice of mathematics as a consequence of the inherent logic of the subject and of the nature of human mathematicians.

Examples? All of successful mathematics! In the geometry of Euclid a concept that deserved a name is that of the right angle, and a theorem used again and again is that of Pythagoras (which uses the concept of the right angle). A more modern example is the concept of analytic function.[4] A theorem used repeatedly is that *a function analytic in a domain reaches its maximum on the boundary*. Professionals may frown at the sloppy statement that I just gave,[5] of the sort used in verbal discussion rather than in writing. In fact, sloppy statements are useful as short formulations of longer theorems "known to everybody." And the practice of mathematics uses a lot of short formulations like *the con-*

tinuous image of a compact set is compact.[6] Sometimes a theorem is only alluded to, and a mathematician will write, "By compactness it is clear that . . . ."

The discussion I have just presented is intended to give some idea of the how and why in the practice of mathematics: unavoidably long proofs, focus on short formulations, and use of appropriate definitions to keep things short. What we get at the end is a *mathematical theory*: a human construct that, unavoidably, uses concepts introduced by definitions. And the concepts evolve in time because mathematical theories have a life of their own. Not only are theorems proved and new concepts named, but at the same time old concepts are reworked and redefined. For the reader who has studied topology, let me mention the emergence of a remarkably useful and natural concept: that of *compact* sets.[7] These occurred first among other classes of sets with somewhat different definitions. Eventually the modern concept of a compact set was singled out as being the "right" one.

I find it very satisfying that we can make sense of concept creation in mathematics as an inevitable consequence of the logical structure of mathematics and of basic features of the human mind. The approach we have taken seems to me preferable to one that would try to understand concept creation *in general*, ignoring the specific features of the substrate (provided here by mathematical logic) and of the human mind (with its deficient memory, etc.).

We must admit, however, that our knowledge of the logical structure of mathematics and of the workings of the human mind remain quite limited, so that we have only partial answers to some questions, while others remain quite open.

One thing we would like to know is to what extent mathematics could have been developed with other concepts than those we are familiar with. In the rock climbing analogy, the question is whether there are several natural routes to the top of a cliff, and the answer is often yes. In mathematics, too, the conceptual structure of a subject can often be developed in different ways. So the reader familiar with measure theory will know that abstract measure theory is favored by some colleagues, while others prefer to deal with Radon measures.[8] And the probabilists (who are a bit isolated in the community of mathematicians) study measures with their own terminology (*marginals, martingales,* etc.) and their own concepts and intuition. Sometimes, a new branch of mathematics is created for extramathematical reasons and turns out to be of remarkable intrinsic interest, or it may illuminate older parts of mathematics. So it was that the emergence of electronic computers led to the development of a theory of algorithms with important new concepts like that of NP completeness,[9] which would have had little chance of being encountered otherwise. I was personally involved in a story that I shall tell in a later chapter. There a concept of a Gibbs state was developed in the mathematical study of a branch of physics called *equilibrium statistical mechanics.* Later Gibbs states turned out to be a remarkable tool in the study of the so-called Anosov diffeomorphisms, although the latter have a priori nothing to do with statistical mechanics. These examples contradict the notion that good mathematical concepts only arise from internal mathematical necessity. Sometimes this is the case, but sometimes concepts coming from outside turn out to be powerful and are eventually considered as natural.

I hesitate to ask the next question: what structure could nonhuman mathematics have? In the rock climbing analogy, it is clear that the problems encountered by a lizard or a fly in going up a cliff are vastly different from those met by a human climber. While it is difficult to imagine nonhuman mathematicians,[10] we have seen from the example of computers that perhaps they could handle some questions better than we can (because of having better memory, working faster, and making fewer mistakes). Also, if you think of it, biology (i.e., natural evolution) gives an example of some kind of nonhuman intelligence: it has solved many difficult engineering problems and, besides that, has created a brain that can do mathematics! Yet evolution works by trial and error in what appears to be a totally *nonconceptual* fashion.[11]

Returning to human mathematics, we have seen why it is necessarily based on concepts or, if you like, structures. But the explicit introduction of structures in the modern sense (as you find them in Bourbaki) is relatively late. For example, the abstract *group* structure appears only in the late eighteenth and nineteenth centuries. Once introduced, these structures turned out to be extraordinarily useful, and they are now essential for many parts of mathematics. But to what extent are these structures inevitable? Are they the natural backbone of mathematics finally revealed in the nineteenth and twentieth centuries? Or are they some kind of scaffolding that is certainly very efficient but basically artificial? (In the rock climbing analogy, think of a metallic ladder allowing you to get to the top of a cliff with minimal effort.)

This question of naturality of mathematical structures, to the extent that it makes sense, probably does not have a simple and clear answer. Look at the subtitle of Bourbaki's treatise: *the fundamental structures of analysis.*[12] Many mathematicians would agree that the structures considered by Bourbaki are natural and may be unavoidable. But a more dynamical view of structures is possible, such as that provided by Grothendieck, which has been described as follows: *one does not make a frontal attack of a problem; one envelops it and dissolves it in a rising tide of very general theories.*[13] While Grothendieck's way of doing mathematics is extremely structural, "hyper-Bourbakist," it does not forget the problems that it wants to solve or dissolve. If Bourbaki's treatise may be viewed as a museum of structures, Grothendieck's effort was the imaginative development of general ideas to understand new and old areas of mathematics. As said before, Grothendieck's program was extraordinarily successful and led to the solution of important problems, generally by other people.

To summarize, we may say that general structures are a remarkable tool for the study of certain parts of mathematics. They are useful and appear natural and even unavoidable to modern human mathematicians, but how deeply natural and unavoidable they really are remains an open question.

We shall later discuss in more detail how we, humans, create new mathematics. In this dynamical view, a good choice of mathematical concepts or structures plays an essential role.

❖ **15** ❖

Turing's Apple

THE JOY OF MATHEMATICAL understanding and discovery is not easy to describe, but it is extraordinary. May I repeat the old story of $\sqrt{2}$? We know that the diagonal d of a square of side 1 is $\sqrt{2}$ (i.e., $d^2 = 1^2 + 1^2 = 2$ by the Pythagorean theorem). Can we write d as the quotient m/n of two integers; that is, can we have $\sqrt{2} = m/n$? No, because this would imply $2n^2 = m^2$. We know that an integer can be written in a unique manner as a product of primes. But the prime 2 occurs an even number of times in the product representing m^2 and an odd number of times in $2n^2$. Therefore, $2n^2$ cannot be equal to m^2.

Clearly, the fact that $\sqrt{2}$ is *irrational* (i.e., cannot be written as m/n) will leave many people deeply indifferent: perhaps the statement and its proof are beyond their intellectual capabilities, or perhaps they simply do not care. But since you have reached the present chapter of this book, you are probably a different kind of person. You see that we cannot restrict ourselves to using numbers that are quotients of integers, and you understand that this is a momentous discovery. This discovery is two and a half thousand years old, and has the beauty of Greek statues but not their fragility. The beauty of mathematics is outside of time. Its multiple treasures are constantly open to the visitor: that not only $\sqrt{2}$ but also the number π is irrational,[1] that finite simple groups can be listed, and that some questions cannot be answered in the conceptual framework of the ZFC axioms. Concerning this last point (Gödel's theorem), one might say that some of the deepest answers to philosophical problems have been obtained in mathematical logic.

You can enjoy some of the beauties of mathematics without being a professional, just as you can enjoy music without playing an instrument or being a composer. But active research in mathematics gives intellectual rewards different from those enjoyed by a spectator. To become a successful research mathematician, you have to be gifted in the first place (as for many other activities), and you also need proper training, luck, and hard work. Specific to mathematics, among other sciences, is that it is a discipline with great freedom, where there are no restricted areas or secret doctrines. You will rarely be asked to show your diplomas, and it is not important that you look intelligent. (There are those who try to look intelligent by frowning, squinting, looking down their nose or up at the ceiling as if to consult with God Almighty; all this is unimportant, just forget it.) Knowing the work of colleagues is important, but it is not team research. (Scientific research outside of mathematics and theoretical physics is essentially done by teams.) So you have a chance to escape the master-slave relations and their ambiguities, which often come with a hierarchical organization. Of course some mathematicians want to be masters, some want to be slaves, and some will try to involve you somehow in their particular neuroses. But you can stay away from them if you are lucky and if you so wish. Mathematical research is a highly individual enterprise. It requires mental agility and the patience to pace around an infinite and dreary logical labyrinth until you find something that has not been understood before you: a new point of view, a new proof, a new theorem.

René Thom once remarked to me that only in mathematics (and perhaps, he added, in theoretical physics) does one find really nontrivial logical thinking. Of course there is very subtle logical thinking involved elsewhere, but not the very long concatenation of rigid logical arguments leading to a statement that can, afterwards, no longer be put in doubt. With mathematics and particularly mathematical logic, we come to grips with the most remote, the most nonhuman objects that the human mind has encountered. And this icy remoteness exerts on some people an irresistible fascination. On what kind of people?

Mathematicians are a very diverse bunch of individuals: men and women of all kinds of ethnic origins, talented or not outside of mathematics, pleasant or unpleasant, with a fine sense of humor or apparently none at all. Their way of doing mathematical research is also diverse (and I leave out of consideration those people connected somehow with math but who say they have, unfortunately, no time left for research). Yet, in spite of all this diversity, there are some features that recur among mathematicians in a statistically significant manner. Indeed, while some abilities are needed or desirable to do mathematics, others are optional, and so it makes sense that mathematicians are statistically different from soccer players, for instance. But one can be both an excellent mathematician an an excellent soccer player, as shown by the example of Harald Bohr.[2] Another reason why mathematicians may be different from other people is that their intense and very abstract activity may eventually have an effect on their health and personality.

The brain is a mathematician's main professional tool, and it has to be kept in reasonably good shape. This excludes the sort of heavy drinking and drug use in which some artists indulge. Of course, many mathematicians drink coffee or tea to stay alert. Tobacco may also help intellectual concentration, although some of its other effects are quite catastrophic. At a time, in the 1960s, marijuana was very widely used in American academia, including among mathematicians, but I heard no claim that it helped their mathematics. A curious remark is that some mathematicians will drink wine to slow themselves down. Indeed some fast-thinking individuals tend to accelerate uncontrollably during complicated arguments or calculations, when one should actually slow down to avoid mistakes. A moderate amount of wine might thus be helpful to some people. In the same vein, I was told by a colleague that, after he had taken codeine for some medical reason, he went with great patience through a long and complicated mathematical argument: he had all the time in the world. In general it is accepted that drugs don't make you more intelligent. Therefore, there is not the sort of drug problem among mathematicians that exists among athletes or some artists. Certainly, there is hedonistic use of wine and other drugs (legal or sometimes illegal), and there is occasional abuse. I think, however, that the main problem worth mentioning with respect to drugs and mathematics is the very painful period which many colleagues went through when they decided to stop smoking, and could not properly concentrate on their work.

Civilized nations strive, in principle, to make humans legally equal. Natural talent, however, and intellectual environment are very unevenly attributed to us. So some people are not good at math, while others move around mathematical problems with the apparent ease and lightness of a dancer on stage. For this, certain gifts (in particular, a good short-term memory) are of course helpful. One may also mention the ability to concentrate or an aptitude for abstract thinking, but these are somewhat fuzzy psychological notions and of limited interest to us as we try to understand mathematical thinking.

When I mentioned earlier escaping a master-slave relation or avoiding involvement in other peoples' neuroses, you may have felt that you handled such problems all the time, and no big deal. This probably means that you are socially well adapted: you communicate easily, your "sexual preferences" are accepted in your community, and so on. Many mathematicians are socially well adapted, but it is interesting that many are not. Why is that? The notion is that if you are intelligent but communicationally deficient, you will turn your interest to activities with limited social demands. These include mathematics, computer programming, and some forms of artistic creation. As an example we may think of the great Kurt Gödel, a person obsessively preoccupied with his health, and with limited social gifts. He had, one may suppose, a very rich inner life, but his relations with the outside world seem to have been mediated largely through his wife, Adele. When Adele was incapacitated with disease, he was left to face his problems alone, in particular the obsession that people were trying to poison his food. Self-inflicted starvation eventually caused his death, sitting on a hospital chair in Princeton, New Jersey.

There is a cluster of conditions known as autism, where communication, social relations, and imagination are impaired. The nature of autism is not understood, but genetic factors are known to be important. It has been argued that "mild autistic traits can provide the single-mindedness and determination which enable people to excel, especially when combined with a high level of intelligence."[3] In fact, Newton, Dirac, and Einstein would be examples of people with Asperger's syndrome, a form of autism. This is an interesting assertion but to be taken with a grain of salt since Newton, Dirac, and Einstein were not medically tested for the syndrome in question. In any case I think there is something peculiar about many (not all) mathematicians: a somewhat rigid way of thinking and behaving. The evidence on which I base this opinion is anecdotal, not clinical. To be specific, my experience is that many mathematicians will give excessive detail when answering a casual question (on the rules of the game of checkers, say, or genealogy in feudal Japan), or they will find logical difficulty with an assertion that causes no problem for most people. Or perhaps they may ask you to repeat a joke and then ask you to explain why it is funny. Let me repeat that not all mathematicians—thank God—are like that. One finds among them a great variety of psychological types and even psychiatric disorders, provided the latter do not impair intelligence. Paranoid, manic-depressive, or obsessive tendencies are not rare among scientists in general, but there are also many who are depressingly normal and dull.

Specific to mathematicians is that, in a professional situation, they have to react in a way that is different from that of most people. If you participate in a public debate or perform delicate surgery, you may have to make quick decisions: some decisions are better and others not so good, but wavering and failing to decide is the worst choice. If you work on a mathematical proof by contrast and you are suddenly not sure that what you say is correct, you should freeze, take your time, and make absolutely certain that your argument is watertight. With one type of personality, you will do wonderfully well as the host of a television talk show, but you would fail miserably as a mathematician. Or you could be a truly great mathematician and look pathetic in a television program.

We have just argued that having a certain type of personality may help one to become a good mathematician. But logically we must also admit that mathematical work may affect one's personality. I think that this is indeed the case, simply because high-level mathematical research is very hard work. The evidence for high-level mathematicians having nervous breakdowns is impressive, if anecdotal. In her biography of David Hilbert,[4] Constance Reid devotes just a few lines to the disappearance of Hilbert for several months in a sanatorium because of a nervous breakdown. She mentions on this occasion the earlier and more severe breakdown of Felix Klein and relates the opinion of Courant that "almost every great scientist I have known has been subject to such deep depressions."[5] One might compare doing great mathematical work with climbing high mountains: they are admirable feats, but dangerous. The mind in one case and the body in the other are pushed to their limits, and there is a price to be paid. Apart from a nervous breakdown, the way mathematicians overuse their brain often results in absentmindedness and lack of practical sense (poets have a similar reputation). And perhaps another result of brain overactivity is baldness, commonly seen among intellectuals (eggheads).

Research mathematicians, then, are doing very hard work but live to some extent in a separate universe and are spared some of the relational problems of "real life." These unresolved problems may, however, come back brutally to the surface and demand attention. The story of the British mathematician Alan Turing[6] is an example of this.

Born in 1912, Turing made his best remembered scientific contribution in the 1930s with the concept of a universal computer, now known as the Turing machine. He gave a precise description of a finite automaton, with infinite memory, which could do any computation that any other such automaton could perform. The idea is that, if you have a suitable digital computer, you can program it to do any calculation that any other computer can perform. Programmable computers did not exist at the time. It was a new idea, which clinched the concept of computability and clarified Gödel's work. Of momentous historical importance was Turing's breaking of the Enigma cipher used by German submarines at the beginning of World War II. This ensured the control of the Atlantic by Allied forces. He also worked on the development of electronic computers and contributed to the debate of whether computers can "think" (the Turing test[7]). Finally, he made a seminal contribution to the understanding of how spatial structures are created (*morphogenesis*) in terms of chemical reactions and diffusion. In a sense, Turing was one of many "original" characters doing dangerous chemical experiments over the kitchen sink (he used potassium cyanide[8]) and pursuing various crazy ideas. But Turing's ideas worked. His contributions to science and our understanding of the world stand out, and there is no way they can now be dismissed and forgotten.

Turing was unpretentious in the way he dressed and interacted with colleagues. Frank Olver[9] remembers him working with a team that made very long numerical computations (on desk calculators) to test an algorithm. Turing had to be fired because he made too many mistakes! To those who met him, he may not have appeared as a very striking person.

Turing, however, was gay, and in England in 1952, this was against the law. He was found out. Having pleaded guilty to an "act of gross indecency," he was given the choice between a prison term or medical treatment. The latter, which he chose, consisted of injections of female hormones for a period of one year. This cure of male homosexuality (according to medical notions of the time) was in effect a chemical castration, reversible, in principle, unlike the compulsory surgical castration practiced at the time in some parts of the United States.[10]

Because the accepted ideas on homosexuality have changed, the hormonal treatment given to Turing may now seem absurd and barbaric. Still, it should be clear that the United Kingdom in the 1950s had nothing to do with Nazi Germany or Soviet Russia. It was a highly civilized nation, where male homosexuality was culturally important in the social upper crust. Turing, unfortunately, had too much of the intellectual rigidity frequently seen among mathematicians and not enough of the hypocrisy frequently seen in the social upper crust. He went through social shame and hormonal treatment better than one might expect. But then one day in June 1954, he was found dead in his bed, poisoned by cyanide, with an apple next to him, from which he had taken several bites. Apparently, he had used the poisoned apple to commit suicide. We would like to understand how he came to this decision. But he left no explanation. He did not answer your questions or mine. The apple was an answer, very final, to his own questions.

❖ 16 ❖

Mathematical Invention: Psychology and Aesthetics

M ANY MATHEMATICIANS have pondered on the psychology of mathematical invention. What does introspection tell us? Henri Poincaré[1] and Jacques Hadamard[2] discuss a remarkable phenomenon that they observed about themselves and that a number of other mathematicians have observed as well. After working for some time on a problem (the *preparation* period) and having failed to crack it, they abandoned it. Then, a day, a week, or several months later (the *incubation* period), suddenly, upon waking up or in the course of a trivial conversation, the solution occurred to them. This *illumination* (as Hadamard calls it) comes without warning and may be along lines that are different from the research done before. The illumination is immediately convincing, although serious checking needs to be performed later. This last stage of *verification* (verifying the solution and making it precise) may show that the illumination was wrong, and then one forgets it. But often enough the solution provided by the *Gods* turns out to be correct. Instead of the *Gods* one now prefers to speak of the *unconscious*. But you may, like many people, be equally unhappy with the unconscious and the Gods. Let me thus proceed with some care.

Consciousness is an introspective concept. When you ride a bicycle or drive a car, you may consciously decide to turn right. But many things that required an effort when you learned riding or driving (like keeping your balance or putting your foot on the brake) you now do automatically: unconscious mental processes are at work. We can thus introspectively recognize conscious mental processes and infer that many other processes take place that are not conscious. Those many unconscious processes appear to be of a very disparate nature, and it is probably misleading to lump them together as *the* unconscious. Also, since consciousness is introspective, it is difficult to define.

How do you know if your spouse, your cat, or your PC has consciousness?

I do not wish at this point to get mired in the general problems of consciousness, the unconscious, the nature of thinking, of understanding, of meaning, the immortality of the soul, and so on. Of course these are interesting problems, but their study entails formidable methodological difficulties. My attitude here will be to ask what one can say about some of these problems in a special but methodologically favorable case: that of mathematical work.

I shall assume that mathematicians (and probably many other people) have, like myself, an introspective notion of consciousness. There is then the interesting claim, by some prominent mathematicians, that an important part of their mathematical work is done unconsciously. We have just seen that Hadamard, following Poincaré, distinguishes in mathematical work a conscious stage of *preparation*, an unconscious stage of elaboration or *incubation*, an *illumination* that reverts to conscious thinking, and a conscious stage of *verification*. The incubation stage is described as *combinatorial* in nature: ideas are put together in various ways until the right combination is chosen. And it is claimed that this choice is made on an *aesthetic* basis. Hadamard sees a mathematical argument as generally consisting of several parts, each of which has a preparation, incubation, illumination, and verification structure. The verification of one part leads to the precise formulation of a *relay result*, which can then be used as a basis for the preparation stage of the next part of the argument. According to many accounts, mathematical thinking is not necessarily based on language. The concepts used can be nonverbal, associated with vague visual, audio, or muscular elements. Hadamard says that he himself thinks in terms of nonverbal concepts and that afterwards he has a hard time converting his thinking into words. Einstein, in a letter to Hadamard, indicates that his own scientific thinking is of a combinatorial nonverbal nature. As for consciousness, he says: "It seems to me that what you call full consciousness is a limiting case which can never be fully accomplished. This seems to me connected to the fact called the narrowness of consciousness (*Enge des Bewusstseins*)."[3]

Where does this leave us? Can one add anything to what great masters like Poincaré, Hadamard, and Einstein have said? I think one can and one should. First, because none of these great scientists defended the *magister dixit* philosophy (i.e., the master has spoken, and this closes the discussion). Second, because the intellectual landscape has changed since Hadamard wrote his wonderful little book. One point that I have raised earlier concerns what we have learned about short- and long-term memory: part of the incubation period probably involves putting into long-term memory the work of the preparation period. This explains why, after some work on a problem (that Hadamard called *preparation*) it is often good to lay it to rest for a while.

An important change in our intellectual landscape came with the advent of powerful digital computers. We now want to compare the performances of the human mind and those of computers, and we naturally ask how we could program a computer to emulate the work of the human mind. From this point of view we have noted that long numerical calculations are carried out easily and without error by a computer, but that translation from one language to another remains difficult. Indeed, a language does not just consist of a dictionary and rules of grammar; it also has many hidden rules and a vast corpus of references that we use to produce a flexible, reasonably unambiguous, and idiomatic output. It is probably significant that the rules of language are difficult to program into a computer but are (in part) needed to do mathematics.

A mathematician who has finally understood a question may say that it was after all very simple. But this is usually an erroneous feeling. In fact, when our mathematician starts writing things down, their complexity unfolds and may end up looking formidable. A simple mathematical argument, like a simple English sentence, often makes sense only against a huge contextual background.

Returning to computers, I like to play with the idea that they could be programmed to invent good new mathematics. And this raises an obvious question: how do we program ourselves to do mathematics? Following professional usage, what I mean by "doing mathematics" is an active, constructive process. Imagining the properties of a mathematical object and trying to prove them is "doing mathematics." For instance, the mathematical object may be a class of dynamical systems, or a theorem about such systems, or a paper you are writing on this topic. Reading a mathematical paper may or may not be "doing mathematics" depending on whether it corresponds to constructing something in your mind. "Doing mathematics" is thus working on the construction of some mathematical object and resembles other creative enterprises of the mind in a scientific or artistic domain. But while the mental exercise of creating mathematics is somehow related to that of creating art, it should remain clear that mathematical objects are very different from the artistic objects that occur in literature, music, or the visual arts.

The idea that artistic creation and creative work in mathematics are somehow related brings us back to Hadamard's statement that good mathematical ideas are selected on an aesthetic basis. In fact Einstein made a similar statement about his own work in mathematical physics. Do we then have to believe that good mathematicians are aesthetically talented in other domains like literature, painting, or music? The answer is negative. Many scientists try their literary talent on an autobiography, others paint or play an instrument. The results are often not bad, but rarely great. And in many cases really good scientists achieve truly mediocre artistic results.[4]

The aesthetic competence for mathematics is thus distinct from artistic competence. Can we analyze aesthetic competence? Are we not reaching here the domain of the unknowable? Actually, I think that aesthetic competence for mathematics is easier to analyze than artistic competence. But let me first point out a change in the intellectual landscape since the days of Poincaré, Hadamard, and Einstein: we have become much more aware that art depends on cultural tradition and that cultural tradition is diverse.

Taste for Bach or Beethoven is an acquired taste, as is taste for good wine or taste for good mathematics. This does not mean that one has to be a professional musician to feel that Bach and Beethoven mastered compositions of impressive size and complexity. But we have this feeling because we are familiar with a certain musical tradition. Exposed to unfamiliar music, we may like it or not, but we cannot say if it is joyful or sad or if it is good or mediocre. Tradition changes, of course, and both Bach and Beethoven changed the course of Western musical tradition.

Much that was just said about music (or art) can also be said about mathematics (or science): one can distinguish different mathematical cultures depending on time and place, and sub-cultures corresponding to different schools, approaches, and areas of mathematics. One can thus distinguish French and Russian traditions, algebraic and geometric styles of doing mathematics. Within a culture or subculture some concepts (like that of the *group structure*) and some facts (like the *implicit function theorem*[5]) are well known. But where is aesthetics? Where are bad and good taste? Since I cannot give the relevant definitions and details, the examples that I shall sketch below may remain a bit vague for nonmathematicians. As for mathematicians they will see what I mean and construct their own detailed examples.

Suppose that you are writing a mathematical paper and that, from some mathematical object a, you are constructing an object b. It may be that there is a group G naturally occurring in your problem, such that b is the inverse a^{-1} of a in G and that this fact is of great help in the construction of $b = a^{-1}$. Not seeing that the group G is, so to say, staring you in the face would be an example of bad taste. An example of good taste would be to prove some difficult theorem by a clever application of the implicit function theorem in a Banach space. The implicit function theorem is fundamental and well known, but you may have to be clever in choosing the Banach space and the function to which you want to apply the theorem. If you succeed, you may get a short proof of what would otherwise be a hard result.[6]

Mathematical good taste, then, consists of using intelligently the concepts and results available in the ambient mathematical culture for the solution of new problems. And the culture evolves because its key concepts and results change, slowly or brutally, to be replaced by new mathematical beacons.

Mathematical aesthetics, while culture dependent, is not meaningless fashion. Remember that short mathematical statements may have very long proofs, but that in normal mathematical practice one tries to use shortcuts, making simple applications of well-known theorems and forgetting about the hard proofs of these theorems. A given mathematical culture at a given time refers to standard theorems, procedures, and ways of thinking that define the culture. So, a contemporary mathematician should know the implicit function theorem and the ergodic theorem and be able to apply them. But note in passing that the ergodic theorem, for example, was not part of the cultural landscape of Henri Poincaré: he died in 1912 while the ergodic theorem dates from 1932.[7]

The intellectual landscape of a given mathematical culture has its standard theorems, terminology, and ideas about which people agree. Instead of an arbitrary fashion, they are an efficient way of doing mathematics. But one must admit that historical accidents play some role in the choice of the standard theorems, of the terminology, and of what is considered interesting research. In this sense fashion does play a role in mathematics.

And let me repeat that, in mathematics as in art, the landscape changes. There are golden periods but also long stretches of dull mediocrity. Some innovations are dead ends, blind alleys. Some innovators shine briefly and then are forgotten. Others change the intellectual landscape in a lasting manner.

The Circle Theorem and an Infinite-
Dimensional Labyrinth

Aₛ ᴀ ᴛᴇᴇɴᴀɢᴇʀ I was exposed to a popular Polish way of singing, which I really liked, with extremely shrill women's voices. Unfortunately, I have not heard this way of singing for many years. Since I do not know Polish, I did not understand the meaning of the songs, but that was not too important: what counted was the distinctive shrill style.

Let me now confront you with a piece of mathematics, in fact, a rather loosely presented mathematical text, chosen because the concepts involved are standard and will be readily understood by mathematicians. As for my nonmathematical readers, they will be (for the length of the next paragraph) in the same position where I am with respect to songs in Polish: able to appreciate, if not the detailed meaning, at least the tune and style of singing.

The beginning of the story is that the physicists T. D. Lee and C. N. Yang, encountered a particular class P of polynomials

$$P(z) = \sum_{j=0}^{m} a_j z^j$$

while studying a problem of statistical mechanics. The polynomials P in P that they could analyze had all their roots on the complex unit circle $\{z : |z| = 1\}$. They conjectured that this was true in general. If they could find a unitary matrix U such that P (z) is the characteristic polynomial of U, that is, P (z) = det (zI − U), then the conjecture would be proved. This is the idea that will occur to any mathematically educated person, but here it does not help. Lee and Yang were good enough mathematicians that they found a proof of their conjecture, but their proof is not easy. Less difficult proofs exist now, due in particular to the work of Taro Asano. To prove the Lee-Yang circle theorem (which will be formulated below) one replaces the polynomial P of degree m in the variable z by a polynomial Q (z_1, . . . , z_m) in m variables, separately of degree 1 in

91

each variable $z_1, \ldots, z_m$. One is interested in the class Q of such polynomials for which $Q(z_1, \ldots, z_m) \neq 0$ whenever $|z_1| < 1, \ldots, |z_m| < 1$. Therefore, if $P(z) = Q(z, \ldots, z)$ and Q is in Q, the roots ξ of P satisfy $|\xi| \geq 1$. (In the case of interest, there is a symmetry $z \to z^{-1}$, so that also $|1/\xi| \geq 1$, and hence $|\xi| = 1$.) Clearly, if $Q(z_1, \ldots, z_m)$ and $\tilde{Q}(z_{m+1}, \ldots, z_{m+n})$ are in the class Q, then also

$$Q(z_1, \ldots, z_m) \, \tilde{Q}(z_{m+1}, \ldots, z_{m+n})$$

is in Q. We describe now a less obvious operation, called Asano contraction, that preserves Q. Write

$$Q(z_1, \ldots, z_m) = Az_j z_k + Bz_j + Cz_k + D,$$

where A, B, C, D are polynomials in the variables $z_1, \ldots, z_m$ except z_j and z_k. Then Asano contraction replaces the two variables z_j, z_k by a single variable z_{jk} so that

$$Az_j z_k + Bz_j + Cz_k + D \to Az_{jk} + D.$$

Starting with a polynomial Q in m variables, we end up with a polynomial in $m-1$ variables, which is again in Q if Q was in Q. (This is an easy exercise: the root of $Az_{jk} + D$ is minus the product of the two roots of $Az^2 + (B + C)z + D$.) One can check that polynomials in two variables of the form

$$z_j z_k + a_{jk}(z_j + z_k) + 1$$

are in Q if a_{jk} is real and $-1 \leq a_{jk} \leq 1$. (Putting the polynomial equal to zero yields a map $z_j \to z_k$, which is an involution sending the inside of the unit circle to the outside.) Taking a product of polynomials as above, making Asano contractions, and putting all variables equal to z, we obtain the Lee-Yang circle theorem: For real $a_{jk} = a_{kj}, -1 \leq a_{jk} \leq 1$, the polynomial

$$P(z) = \sum_{X \subset \{1, \ldots, m\}} z^{|X|} \prod_{j \in X} \prod_{k \notin X} a_{jk}$$

has all its roots on the unit circle.[1]

The above presentation is not very difficult mathematics, but for a professional mathematician it will probably be a refreshing and welcome change from considerations *about* mathematics: this *is* mathematics. Note that I have only sketched the details of the proof, because the reader is assumed to have enough technical background to complete them (or just say, "Yes, of course"). The assumed technical background (or cultural tradition) contains, in particular, a theorem about the characteristic polynomial of a unitary matrix (mentioned but not needed) and the fundamental theorem of algebra[2] (needed but not mentioned). We are far from a formal deduction based on the ZFC axioms. It would, however, be easy (for a professional mathematician) to give a much more formal presentation. And the idea is that any detail of this more formal presentation could, in principle, be expanded to a fully formal text. Therefore, in principle, the above statement and proof of the Lee-Yang circle theorem could be written as a fully formal text that could be checked mechanically. I believe that such texts will eventually be written and checked by computer. This seems to me the only way to fight errors in proofs, which are becoming a daunting problem for the future of mathematics. But let us leave this question for a later chapter.

A fully formal proof of the Lee-Yang circle theorem would be very long and quite unreadable and uncheckable by a human mathematician. One could say that human mathematics is a sort of dance around such a formal text: one gives a convincing argument that it could be written, but one does not write it. What, then, is the status of the text that I have presented above? It is a piece of human mathematics, allowing a human reader to be quickly and efficiently convinced of the correctness (i.e., formalizability) of a certain deduction; it deals with ideas rather than with formal statements.

What is an idea? Or, more specifically, what is a mathematical idea? Trying to be pragmatic rather than profound, I would say that an idea is a short statement in human mathematical language that can be used in a human mathematical proof. (The statement may be a conjecture or a comment.) As an example I want to identify the main ideas in the above mathematical paragraph on Lee-Yang; I see three of them. The first is the conjecture of a theorem (polynomials of a certain form have their zeros on the unit circle). The second is to replace a statement about the polynomial $P(z)$ by a statement about the polynomial $Q(z_1, \ldots, z_n)$. These first two ideas are due to Lee and Yang. The third idea is that of Asano contraction (due to Asano). All three ideas are unobvious. (The second one replaces the obvious idea of expressing $P(z)$ as a characteristic polynomial.) All three ideas can be expressed succinctly. Indeed, after a few minutes of explanations, a working mathematician can start writing down a proof of the Lee-Yang circle theorem. By contrast, guessing the theorem or finding a proof from scratch is definitely hard work. I have stated three main ideas. Secondary ideas can be automatically interpolated by a professional mathematician.

I shall come back in a moment to the question of how a theorem can have a proof that, although simple, is difficult to find. First, however, I want to ask how it is that the Lee-Yang theorem has a simple proof. We have seen, following Gödel, how certain theorems with a short formulation can have a long proof. We are thus not astonished that the Lee-Yang theorem should be difficult to prove, but we may wonder how a simple proof is ever found. Here is the reason: we have at our disposal a number of results with long proofs that we do not have to prove again. (An example is the fundamental theorem of algebra mentioned earlier.) The cultural background of present-day mathematics contains technical tools that allow us to handle efficiently a great variety of problems. (Our panoply of technical tools results from the selection of efficient tools by our cultural evolution.) A simple proof of the Lee-Yang theorem is thus not a short proof starting from the ZFC axioms; it is a short proof starting from standard (in this case, "elementary") tools of algebra.

The set of tools available to a mathematician may be compared to the system of highways available to a traveler: both provide the means to go efficiently from A to B. But there is an important difference: the choice of an efficient itinerary using highways is usually an easy matter; this is not so for the choice of an efficient mathematical itinerary to prove a theorem. Let me pursue for a minute the analogy between highway system and mathematical panoply. The highway system reflects the geography of a country, which we also know by other methods, so that building another road will not significantly change our knowledge of the geography. The panoply of technical tools of mathematics reflects the inside structure of mathematics and is basically all we know about this inside structure, so that building a new theory may change the way we understand the structural relations of different parts of mathematics.

Let me now go back to asking why it may be hard to find the proof of a theorem even if, in the end, the proof is relatively simple. What it boils down to is that discovering something may be hard, but verifying the discovery may be easy. For instance, discovering your boss's computer password may be hard, but it is easy to use once you know it.

I want to digress on passwords for a minute. Assume that the password of your boss has length 7 (i.e., it is a sequence of 7 symbols), and let us say that there is a choice of 62 values for each symbol, $a, \ldots, z, A, \ldots, Z, 0, \ldots, 9$. Then, the number of possible passwords is $62 \times \cdots \times 62 = (62)^7$. It has 7 as an exponent; that is, it grows exponentially fast with the size of the password (the size, or length, is here taken to be 7). Instead of searching for a password, let us search for an intersection in an American city (where streets are perpendicular to avenues). If we consider a rectangle of size 7 (seven streets and seven avenues) there are only $7 \times 7 = 7^2$ intersections. The number of intersections grows like the square of the size of the region searched, which is much less fast than the exponential growth found for passwords. This is because our search for intersections is two-dimensional. A search for windows would be three-dimensional. (Say there are 40 windows for each floor in a block. Then for 15-floor buildings in a 15×15 block area, there are $40 \times (15)^3$ windows.) A search for a needle in a haystack is also three-dimensional. Searching for an address along a street (say, 10 Downing Street) is one-dimensional. What is the dimensionality of the search for passwords? It is larger than $1, 2, 3, \ldots$, and we may say that it is infinite.

It is time to return to the task of a human working mathematician. This task is an approximation to the task of writing a fully formalized mathematical text, but it is not a close approximation. A human mathematician works with "ideas," of which we have given some examples above. A suitable sequence of ideas will give a proof of an interesting theorem. This is the combinatorial task described by Poincaré and Hadamard: putting ideas together until the right combination is found. How hard is this task? It is not a search in one, two, or three dimensions. It is more like trying to guess a password; it is an infinite-dimensional search. But there is a difference. Unlike symbols for a password, mathematical ideas cannot be put together arbitrarily; they have to fit. (An example is provided by the idea of using the Pythagorean theorem. This is a fine mathematical idea, but it works only in a geometric situation, with a triangle that has a right angle. If you don't have that situation, you can't use the idea. Unless, of course, you introduce geometry and the triangle in your problem. Achieving this, however, will require some new ideas.) Putting together a sequence of mathematical ideas is like taking a walk in infinite dimension, going from one idea to the next. And the fact that the ideas have to fit together means that each stage in your walk presents you with a new variety of possibilities, among which you have to choose. You are in a labyrinth, an infinite-dimensional labyrinth.

I have just described human mathematics as a labyrinth of ideas, through which the mathematician wanders, in search of the proof of a theorem. The ideas are human, and they belong to a human mathematical culture, but they are also very much constrained by the logical structure of the subject. The infinite labyrinth of mathematics has thus the dual character of human construction and logical necessity. And this endows the labyrinth with a strange beauty. It reflects the internal structure of mathematics and is, in fact, the only thing we know about this internal structure. But only through a long search of the labyrinth do we come to appreciate its beauty; only through long study do we come to taste fully the subtle and powerful aesthetic appeal of mathematical theories.

❖ 18 ❖

Mistake!

The landscape of mathematics has a historical dimension. New theorems are proved, and better tools are devised for handling all kinds of questions. At the same time, the problems that remain unsolved become progressively more difficult. I once had a chance to chat about this changing world with Shiing-shen Chern,[1] one of the great figures of twentieth-century geometry. And he explained to me how, when he started his mathematical career, he read the work of Heinz Hopf on fibrations of spheres[2] and found himself at the frontier of the mathematics of the time; he could start doing his own original work. Now, Hopf's ideas are wonderful but relatively easy to study. At the beginning of the twenty-first century, it is typically much harder to get to the frontier of mathematics. Think of having to master Grothendieck's ideas, among others, if you want to work in algebraic geometry and arithmetic!

Mathematics does not always become more difficult as time goes by. Sometimes a new technical development provides access to questions that had hitherto been out of reach. Sometimes problems that had not attracted interest become the center of a bright new field of mathematics, with important results relatively easy to obtain. For instance, the arrival of fast computers promoted the study of algorithms and led to basic conceptual developments like the notion of NP completeness and the remarkable proof that primality can be tested in polynomial time.[3]

In general, however, one must admit that mathematics becomes more and more difficult with time. This causes changes in the practice of research. I remember hearing criticism in the 1960s addressed to a mathematician who used results by others without sometimes checking them personally. Because of the inflation of the literature, this checking of earlier results is less and less possible. I heard Pierre Deligne, in the 1970s, stating that the mathematics that interested him was that which he could personally understand in complete detail. This excluded, he said, proofs using computers and extremely long proofs that a single person could not dominate. In fact, however, computer-aided proofs and extremely long proofs have become a normal feature of contemporary mathematics.

Perhaps we are witnessing a decay of the "moral values" of mathematics; Grothendieck said so explicitly. At the same time, however, we see extraordinary successes in the solution of old problems (Fermat's last theorem, the Poincaré conjecture,[4] etc.), and we must admit that contemporary mathematics is in a sense remarkably healthy. Simply, we observe that the nature of human mathematics is changing. And different people adapt to the change in different ways. An example that has led to some controversy arose from William Thurston's work on three-dimensional manifolds. A natural problem of geometry is to *classify* manifolds of a certain kind (this means listing them). The classification of two-dimensional manifolds is well understood. But the study of three-dimensional manifolds is much harder. After considerable work, Thurston came to a good understanding of the subject, of which he gave a broad description, with outlines of proofs. Thurston's program thus laid claim to a big mathematical area but without providing proofs that colleagues could check. In effect, he made it difficult for other mathematicians to work in this area: you don't get much credit for the proof of a theorem that has already been announced, but at the same time you can't use this theorem, because a proof doesn't quite exist yet. A much-discussed paper by Arthur Jaffe and Frank Quinn,[5] mentioning Thurston and others, complained about this evolution of mathematics. As it turns out, Thurston's program has now been largely implemented, but the problem raised by Jaffe and Quinn remains significant for some parts of mathematics.

It is now time for us to look into the mathematical use of computers. Speaking of computers one thinks of long numerical calculations. Are such calculations useful in pure mathematics? Sometimes they are. In fact, Riemann did long numerical calculations by hand to test some ideas and would surely have been pleased if he had had a fast computer at his disposal. Computers have also been of great help in visualizing objects that occur in the theory of dynamical systems.[6] There is thus no doubt that computers can be of use in the *heuristics* of mathematical problems, that is, they make some conjectures plausible and invalidate others. Most mathematicians have no objection to this heuristic role of computers. But the normal use of computers gives only *approximate* numerical results; how can they be used to obtain *rigorous proofs*?

Computers are really rather versatile machines. Let me mention some tasks that they can perform exactly, which can be of use in proving theorems. Exact calculations with integers is the most obvious thing. Computers, however, can also be programmed to do logical operations: checking, for instance, a large number of situations and giving in each case a yes or no answer to some question. This *combinatorial* capability of computers is what was put to use in the proof of the four-color theorem.[7] Computers can also handle real numbers like π or $\sqrt{2}$ exactly, by using *interval arithmetic*. The idea is that, if you know that π is in the interval (3.14159, 3.14160) and $\sqrt{2}$ in the interval (1,41421, 1.41422), you also know that $\pi + \sqrt{2}$ is in the interval (4.55580, 4.55582) *without any error*. Interval arithmetic allows one to perform, with strictly controlled accuracy, all kinds of calculations involving real numbers. Let me sketch an example of how such calculations can be used to prove a theorem. Suppose we know that two (explicitly specified) curves A_0 and B_0 in the plane intersect at a known point X_0, and we want to prove that the (explicitly specified) curves A and B intersect at a point X close to X_0.

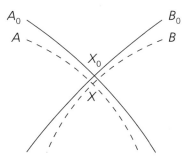

This is known to be true under certain conditions (transversality of the intersection of A_0 and B_0, closeness of A to A_0 and B to B_0 in a certain sense) that can be checked numerically. It may be convenient to do the numerical checking with the help of a computer. I have just outlined a computer-aided proof that, under suitable conditions, two curves A and B have a point of intersection X, with an estimate of the distance of X to a known point X_0.

As it happens, some theorems of real mathematical interest are of the form just described, but with curves A and B replaced by manifolds in some infinite-dimensional space. My colleague Oscar Lanford reported once on a theorem of that kind.[8] The contents of the theorem will not concern us here. Rather, we shall look into some technical aspects of how Lanford proved it. The proof is computer aided, which means that it consists of some mathematical preliminaries and then a computer program. This program (or code) uses interval arithmetic to check various inequalities; if these are found to be correct, the theorem is proved. The complications of the problem forced Lanford to write a relatively long program, about 200 pages. The pages consist of two columns; one has the code (in a variant of the C programming language), and the other has explanations of what one is doing. Indeed, long code without explanations is incomprehensible, even to the person who wrote it. And in the present case, since it is a mathematical proof, other people should be able to check it. Oscar Lanford is a very careful person, and he took pains to check that, when the code is fed into the computer, the computer does exactly what it is supposed to do. In this manner—after the computer has agreed with the inequalities in the code—the proof of the theorem is complete.

But Lanford added some remarks that you may find rather disheartening. "I am sure," he said, "that there are some mistakes in the code I wrote. But I am also sure that they can be fixed and that the result is correct." What this means is that in 200 pages of text, there are probably some mistakes. In the present case it might happen that some inequality that needed to be proved was, in fact, not proved! But Lanford believes that he has a sufficiently detailed understanding of the problem at hand and that he could find and prove a similar inequality, sufficient to establish his theorem.

It is good to remember at this point that computer-aided proofs are not completely formalized mathematics (that one could, in principle, trust completely). Computer-aided proofs are part of human mathematics. The problem of avoiding mistakes when you use a computer is different, however, from what it is in "normal" pencil-and-paper mathematics. You can check and flush out some kinds of mistakes present in computer code, but you don't have the intuition that good professional mathematicians have developed about pencil-and-paper proofs.

With proofs getting longer and longer, the problem of mistakes in mathematics is becoming increasingly serious with time, whether or not computers are used. Together with mistakes, I shall here discuss gaps, that is, elements of a proof that are supposed to be easy to see, but aren't. To put it bluntly, the probability that there is no mistake in a proof goes down with its length in an exponential manner (or worse). And a single mistake can kill a proof! It is fortunate that many mistakes, like misspelling a name or putting a wrong date in a reference, are mathematically inconsequential (even if they can make some people quite furious). More serious mistakes can often be fixed too, and we shall later discuss how that happens. We can see how grave the problem of mistakes or gaps can be by looking again at the theorem of classification of finite simple groups. The proof of this theorem covers many thousands of pages, by many authors, and parts of the proof are computer aided. The theorem has been considered as "morally" proved since around 1980, with some parts yet to be written. This means that there were gaps in the proof, but they were not considered serious by specialists. One of those gaps, however, turned out to be serious enough to necessitate another 1,200 pages of proof (in 2004[9]). There are other areas of mathematics that are in a messy state. For instance, speaking of the packing of spheres, Tom Hales wrote, "the subject is littered with faulty arguments and abandoned methods."[10]

Does this mean that mathematics has forgotten its old standards of rigor? That mathematical truth has become a matter of opinion rather than a matter of knowledge? Interesting views on this problem have been expressed by various authors in response to the article of Quinn and Jaffe mentioned earlier.[11] Basically, one may say that good mathematicians working in an area know how reliable the published literature is. Some areas have been scrutinized repeatedly by high-level mathematicians and theorems proved by different methods; such areas can be considered as extremely rigorous. But one must admit that the mathematical literature contains lots of junk, because some people need to publish for career reasons even if they have little interest in what they are doing.

In brief, the old ideals of absolute logical rigor have not been abandoned. But there are forces at work to change the style of mathematics, because highly desirable theorems may require very long proofs or computer-aided proofs. Think, for instance, that if you want to prove a property of simple finite groups, you can do it by checking the property in question on an explicit list of groups. This shows how useful the classification theorem is: it is a new beacon that changes the landscape of mathematics. Of course there are changes also for the human mathematician. Being a mathematician today is not what it was a hundred years ago. Doing mathematics in a hundred years will again be different. Perhaps it will be a less satisfying enterprise than it was in earlier centuries, and perhaps not. But there will be new results, deeper theories. And more of the unknown face of mathematical reality will have come to the light of human understanding.

❖ 19 ❖

The Smile of *Mona Lisa*

In the course of a scientific career, one attends many many technical talks. Anyone with experience of this sort of thing will know that occasionally one stops listening. Either one is not interested, or one lacks some required knowledge, or one has missed something important that the speaker said at the beginning of the lecture or should have said. And then one sits, half dozing, lost in thoughts unrelated to the topic of the talk and catching occasionally a technical phrase or a meaningless sentence as it flies by. On one such occasion my attention was gotten by the word *KILL* that came back repeatedly and with a strong stress. In fact, *KILL* came in the phrase "KILL that antisymmetric matrix." Technically, a matrix is a table (a_{ij}) of numbers, it is antisymmetric if $a_{ij} = -a_{ji}$, and "killing" the matrix may have meant "finding an eigenvector with eigenvalue 0," but I am not sure. There was, however, something funny in the way the speaker pronounced "KILL that antisymmetric matrix." What he actually said was, "KILL that anti-Semitic matrix." I was wide awake now, my hearing was quite good at the time, and I listened carefully. If I had paid attention to the mathematics, I would have missed it, but there was no doubt: he said *"anti-Semitic,"* not *"antisymmetric."* Repeatedly he said, "KILL that anti-Semitic matrix."[1]

Of course the words *kill* and *matrix* may have a mathematical meaning, but they also have a meaning in everyday English: *kill* meaning "to kill" and *matrix* meaning "womb." These profane meanings are occulted, repressed, in a mathematical discussion, but they are still there at some unconscious level, as the above story shows. We have discussed earlier the nice aseptic unconscious that provided Poincaré and Hadamard with solutions to their mathematical problems. But "KILL that anti-Semitic matrix" is eruption of another kind of unconscious, loaded with sex and unpleasantness: the unconscious of Dr Sigmund Freud. Need we really go into that? Can Freud's ideas bring something useful to our discussion of mathematics? I invite you to hear the case and then make your own judgment.

I am not suggesting here that Freud's ideas are the key to understanding the nature of mathematical thought. My conclusion will be that they are not, and Freud in fact makes no claim of that sort. This being said, would it not be nice to interview Freud and ask him why, in his opinion, some people engage themselves in mathematical work? While Dr. Freud is no longer with us, we can have a look at his *Leonardo da Vinci and a Memory of His Childhood* (1910).[2] This study is indeed quite relevant to a discussion of the origins of scientific curiosity.

The Florentine Leonardo da Vinci (1452–1519) is of course the painter of the *Last Supper, Mona Lisa,* and a few other masterpieces. The notebooks that he has left record an insatiable scientific curiosity in observing nature and an amazing mechanical inventiveness. He was centuries ahead of his time intellectually. It is no wonder that such a personality would attract and captivate the attention of Freud.

Leonardo was the illegitimate son of Piero da Vinci, a Florentine notary, and Catarina, a young peasant woman. At age five, Leonardo was in the household of Ser Piero da Vinci, the latter now married to Donna Albiera, who remained childless. Around the age of fifteen Leonardo became apprenticed to Andrea del Verrocchio and turned into the extraordinary artist that we know. Later he spent more and more time in the studies described in his notebooks: of nature, engineering and other topics.

Freud points to remarkable features of Leonardo's personality, some of which beg for an explanation. He was a strong and handsome man, who liked to dress elegantly and live in good company. Probably he had some homosexual tendencies, but no real sex life.[3] Leonardo left paintings unfinished after years of slow work. His appetite for understanding was immense, and work on his notebooks progressively displaced his painting activity. He planned several treatises but was unable to finish any. He was apparently a vegetarian, condemned war, and bought birds at the market, which he then let fly away. But he also attended the execution of criminals and served as chief military engineer of Cesare Borgia.

Perhaps one should add that the "science" of Leonardo was turned to the visual description of nature and was thus directly related to his painting. He studied perspective, dissected dead bodies, and observed the flight of birds. His modernity is revealed by statements such as *"Referring to authority in a discussion is using one's memory rather than one's intelligence,"* and also *"Nature is full of multiple reasons that were never reached by experience."*[4]

Before we look into Freud's analysis of Leonardo, let me mention that Newton shared some of the characteristics discussed above: the same immense appetite for understanding and diverse interests (unified by a desire to discover the nature of the universe), and some homosexual tendencies, but an apparent lack of sex life.[5]

Freud explains Leonardo's personality in terms of *sublimation*. According to Freud, sublimation is a process by which the sex drive serves as a motor for some activities apparently unrelated to sexuality: artistic activity and intellectual investigation.[6] Young children are naturally faced with the ontological problem of their own origin: how are babies born? The standard answer in Freud's day involved storks. But an intelligent child might guess that the role of its mother was more important than that of storks and would face some formidably challenging intellectual questions: What was the real role of the mother? Why did the parents lie about it? What was the role of the father, if any? What is the difference between boys and girls? Why? Seen this way, sexual curiosity, motored by sex drive, appears central to the curiosity of young children. In the normal course of events this curiosity would, in due time, be one of the elements leading to "normal" sexual behavior. (In older days one would have written "normal" without quotation marks.) But part of this curiosity is *sublimated* to nonsexual purposes that may have social value, in particular, artistic activity or intellectual investigation. In some cases, as that of Leonardo according to Freud, the original sex drive is totally converted to nonsexual purposes.

While Freud's ideas in general have drawn quite a bit of opposition, the concept of sublimation has been relatively well accepted. For example, the *American Heritage Dictionary* briefly describes the concept (see *sublimate*), even if Freud is not mentioned. Where we may reproach Freud is for his excessive faith in the powers of the "psychoanalytic method" in a situation, like that of Leonardo, where not enough data is known. For example, Freud remarks that in notes made by Leonardo concerning a pupil to whom he seems to have been attached or about the deaths of his mother, Catarina, and father, Piero da Vinci, numbers are mentioned (like the cost of candles) but no emotion expressed. This is a perceptive remark but it is weakened by the fact that one is not sure if the Catarina referred to here is his mother or just a servant.

The memory of Leonardo's childhood that Freud wants to interpret comes in relation to the study of the flight of a bird (called in Italian *nibbio*). Leonardo points out that it was apparently his fate to write in such detail about the nibbio, because the first thing he remembers from his childhood appears to be that, while he was in his crib, a nibbio came to him and opened his mouth with its tail and struck him repeatedly inside the lips with its tail. To a modern open-minded educated reader, one century after Freud, this "memory" (or fantasy) readily suggests a sexual interpretation. According to Freud, it is an oral-sexual fantasy related to the breastfeeding of baby Leonardo by his mother. Freud read the "childhood memory" in a German translation, where, unfortunately, *nibbio* was translated as *vulture* (*Geier*) when it should be *kite*. As a result he makes a lot of the fact that the word *mother* in ancient Egyptian is represented by the picture of a vulture, and he gets lost in meaningless interpretations based on a mistaken relation between *nibbio* and *mother*.

Freud also interprets the famous painting *Virgin and Child with St Anne* (Anna Metterza) as representing Leonardo (as child) with his two mothers (Catarina and Albiera). In support of this interpretation, Freud remarks that the pictorial idea of the Anna Metterza was unusual when Leonardo did his painting. Unfortunately, as pointed out by the art historian Meyer Schapiro, Freud's remark is in error: the cult of St. Anne and the theme of the Anna Metterza were flourishing at the time when Leonardo worked.

How do you, reader, react to the above discussion? Many of my colleagues, who work in "hard" sciences like mathematics and physics, react to Freudian psychoanalysis and other "soft" bodies of knowledge (like philosophy and economics) in a dismissive way. They will formulate an utterly devastating (and absolutely correct) judgment on how the matter has been handled by so-called experts. And then they may explain what should really be done to solve, for example, the problems of the economy. At that point they will fall into one of the many pitfalls of the subject, well known to the experts.

"Soft" bodies of knowledge are those that are methodologically difficult and insecure. This is certainly the case of Freudian psychoanalysis. Emphatically: Freud is not infallible. But he has unveiled many important concepts. His ideas have had an enormous influence on twentieth-century Western culture, and this includes an unrecognized influence on the ways of thinking of people who don't want to hear about Freud. Some Freudian concepts have become unavoidable. Sublimation is one of them, and it does help us to understand the personality of Leonardo da Vinci and Newton. But, as Freud explicitly recognizes, psychoanalysis does not explain the secret of the smile of the *Mona Lisa*. And neither does it explain, I think, the secrets of mathematical thinking.

Since our interest is in mathematical thinking, why did I bring in Sigmund Freud? Well, so that one does not forget that the mathematician's brain contains many objects: theorems, lemmas, money preoccupations, and also "KILL that anti-Semitic matrix." All these things coexist and interact in obscure ways. Fortunately, mathematical thinking can be logically separated from the rest, and this is what we are doing in the present book. This separation has a great methodological advantage: it isolates an area that can be analyzed in remarkable depth, much better than questions involving psychology. The possibility of analyzing mathematical thinking in a deep way makes the subject of considerable philosophical interest. But one should not forget that, besides beautiful mathematical ideas, there are many more obscure things that crawl in the mind of a mathematician.

❖ 20 ❖

Tinkering and the Construction
of Mathematical Theories

Doing mathematics is often an individual and solitary enterprise. But mathematics as a whole is a collective achievement. A mathematician lives in an intellectual landscape of definitions, methods, and results, and has greater or lesser knowledge of this landscape. With this knowledge, new mathematics is produced, and this invention changes more or less significantly the existing landscape of mathematics. How is this done? What is the strategy of mathematical invention?

One thing is clear: you do not try to obtain systematically all the valid consequences of the ZFC axioms, using formal language and allowed rules of deduction. And neither do you try to obtain the shortest proof of a theorem starting from ZFC and using formal language. You always work in a context, or landscape, of already proved results. In principle, you should be able to translate what you do in formal language, but you prefer to use a natural human language like German, French, or English, which is better at conveying the meaning of mathematical ideas and formulating the aims of your work. Meaning! Aims! Oh those dangerous words! Earlier we discussed mathematical *structures* and mathematical *ideas*. These things are not contained in the axioms, but we have been able to relate them to formal mathematics. *Meaning* and *aims* are another matter. They may be important to discuss the strategy of mathematical invention, but these concepts are—at this point, at least—totally outside mathematics.

We are not, however, trying to define meaning and aims in general but only in the special and reasonably well controlled context of mathematical work. Leaving *meaning* for later, let me concentrate on *aims*.

One may say that the aim of a mathematician at work is always to develop a mathematical theory. Sometimes this is guided work: studying what other mathematicians have done.

Sometimes it is original work. Instead of discussing what a mathematician's aim is, I shall thus describe what the mathematician actually does: constructing a theory. A mathematical theory is a piece of mathematical text, as described in an earlier chapter. Specifically, it is a collection of statements that are logically connected. We may also say that a theory is a coherent construction consisting of mathematical ideas. Perhaps one theorem in this construction is considered more important than the rest, and it will then be said that the aim of the work was to prove this theorem.

The aim of mathematical work, then, is to perform a construction: the construction of a mathematical theory, that is, a coherent collection of mathematical ideas. Of course, one wants the theory to be *interesting*. A theory is interesting if it contains results not known before, preferably with a short formulation and a *nontrivial* proof (i.e., starting from known results, the proof is necessarily either long or unobvious). For a theory to be interesting, it is also desirable that it can be used to prove further results. Interesting mathematical work is judged against the background of a certain mathematical landscape. What is considered interesting is motivated in part by the history and sociology of the subject. But it would be a mistake to reduce the question of the interest of a mathematical theory to a matter of sociology: the logical structure of the theory plays a more basic role. In a given area of mathematics there are usually *conjectures* left unproved by earlier students of the subject and which may constitute a guide to interesting topics. I shall assume that the working mathematician whom we are observing has some definite ideas as to what is interesting. (And we must admit that some mathematicians have, in this respect, better taste than others.)

After a lot of preliminary considerations we have finally reached the central problem of creative mathematics: how do you construct an interesting theory? In practice it may mean this: how do you write a twenty-page paper that will be published in the *Annals of Mathematics* and ensure that you get tenure at a good university? (The *Annals* is a good journal, rather choosy as to which papers it accepts, and it publishes interesting articles in general.) The number of interesting twenty-page papers that are conceivable is quite enormous, and the number of twenty-page papers that are uninteresting, wrong, or meaningless is even more enormous. Trying to write an interesting twenty-page paper confronts us with a problem that we described earlier as finding one's way through an infinite-dimensional labyrinth.

Let us forget about twenty-page mathematical papers for the moment and look more generally at sequences of symbols (mathematical or otherwise) of a certain length. We suppose that a certain *interest* is associated with each sequence, and we want to address questions like: What is the average interest of a sequence? How do I find a sequence of large interest? What is a sequence with maximum interest? Physics, engineering, and financial mathematics give rise to such problems, which one tries to handle using a computer. How does one proceed? There are many methods depending on the specific problem at hand but I would say that there are two basic ideas to keep in mind: (1) using random choices, (2) tinkering.

We first discuss the use of random choices. The number of all sequences of symbols to be considered is usually so enormous that looking at them one by one is hopeless. So, in order to estimate the average interest of a sequence, one does not look at all of them, but one takes a sample. This means that one takes at random 1,000 or 1 million sequences, and computes the average interest of those. This is the principle of what physicists call the *Monte Carlo method* (an allusion to the randomness involved in the casino games in Monte Carlo). One may sometimes improve on purely random sampling, but sampling in any regular way is usually a mistake.

The problem of finding a sequence with strictly maximum interest is normally hopeless, but one can search for a sequence of high interest: look at a number of randomly chosen sequences, and then select the best. One can improve on this method by taking advantage of a feature of many problems: sequences close to one of high interest have higher than average interest. This leads to new strategies, where one makes a random walk among sequences of symbols, with small consecutive steps, and a bias towards more interesting sequences.[1]

The idea of a random walk with a bias towards increasing interest leads to the concept of tinkering, put forward by the biologist François Jacob[2] in connection with biological evolution. Jacob considers among other things the evolution of proteins, and we shall now briefly digress on this question. (Note that, like many other things in biology, the study of protein evolution has exploded since Jacob's 1977 paper.) A medium-size protein is coded by a sequence of about 1,000 symbols, each of which can take four values (the four bases represented by A, T, G, C). There are more than 10^{600} such sequences! An interesting sequence is one that codes for a useful protein (in a given species). Does one find a new interesting sequence by reviewing 10^{600} possibilities? No, it is done by tinkering with already existing sequences. For many protein sequences, an evolutionary history can be traced back 1 or 2 billion years (before that there were chemical evolution and early self-replicating systems, currently beyond reach). A number of protein families have been studied, having a common ancestor sequence, from which they have evolved by local mutations. This is an example of the strategy described above, of making a random walk (one mutation at a time) among sequences of symbols, with a bias towards sequences of increasing interest. The proteins in a given family have the same general shape and may occur in different species, or several different proteins of the same family may occur in the same species. Two proteins of the same family may or may not serve related purposes. What happens is that a sequence coding for some protein may (as a consequence of gene duplication) become available for other purposes. The pressure of evolution may remove the duplicated gene because it is useless, or by mutation, this gene may start a new life and code for a protein with a new use. This means that a new useful protein has been obtained by tinkering with an old one. Tinkering in protein evolution may go beyond local changes in existing sequences. Sometimes pieces of two genes coding for different proteins get joined and code for a new protein. If the *mosaic protein* thus obtained finds some use, it is the first member of a new family, with a shape different from that of the parent proteins.

François Jacob describes biological evolution as a general tinkering process. This process may produce new useful proteins from proteins that already exist, or it may make a wing with a leg or a part of an ear with a piece of jaw and so on. The process of tinkering in biological evolution may be called unintelligent, but it is extraordinarily successful. A human inventor would not be able to design such marvelous products of evolution as a mosquito or a human brain. Note, however, that a human inventor might avoid certain things done by evolution and that appear stupid (like crossing the passage of our food from mouth to stomach with the passage of air from nose to lungs).[3]

It is a natural idea (developed by Aharon Kantorovich[4]) that tinkering, apart from biological evolution, also plays a role in scientific discovery. This is, in particular, the case for the construction of mathematical theories, where one may try random changes of existing concepts in the hope of finding something of interest. Or one may put together in various ways the facts that one knows until one gets a valuable result. This is the hooking up of ideas, which may occur unconsciously, and was described for us by Henri Poincaré and Jacques Hadamard.

But, of course, combining ideas at random is only part of the story. A mathematician working in a certain area of mathematics has definite ideas as to the structures that play a role in this area and will, in part, proceed quite systematically on the basis of these structural ideas. In other words, to a working mathematician, mathematics is a meaningful subject. The meaning has to be discovered. It is not evident but it is there. And now we have to face a serious problem: what sense can we give to the word *meaning* in mathematics?

❖ 21 ❖

The Strategy of Mathematical Invention

IF YOU ARE SOMEHOW connected with a university, you may have been in the library of the math department. If not, I suggest that you pay it a visit. What you will see there are tables with students or faculty at work, some computer terminals, stacks of books, more stacks of past issues of mathematical journals in bound volumes, and also display cases with unbound recent issues of the same journals. Pick one of the recent issues of the *Annals of Mathematics*, *Inventiones Mathematicae*, or any of dozens of other journals. As you leaf through the issue, you will find longer and shorter articles on various esoteric questions. Each article starts with title, name and affiliation of the author, abstract (a short summary), then comes the main text with theorems, proofs, and so on, and at the end of the article are listed references to other articles by various authors. The journal issue in your hands will often have shorter articles with titles containing a Latin word like *errata*, *addenda*, or *corrigenda*. These errata are by authors of papers published earlier, acknowledging that there was something not quite right in their paper and trying to fix it. It may just be that some references had to be added, as "kindly" pointed out by a colleague. But more commonly, the colleague kindly pointed out a real mistake in the proof. Sometimes, then, authors have to admit that their "main theorem" remains unproved, and they propose perhaps a weaker, less interesting result. This honorable defeat is, however, not the typical situation. Mostly, the authors thank the colleague who kindly provided a counterexample to a lemma in their paper, but then point out that the main result of the paper follows from a weaker lemma, which is unquestionably correct.

How is it that, so commonly, an error is found in a paper, but the error can be fixed more or less easily? The answer is that the way the results of a paper are presented is not the way these results were obtained. A paper is a description of a mathematical theory (or piece of theory) constructed by the author. The construction involves guessing various mathematical ideas and their relations. The ideas are often problematic (something appears obvious but should be checked later, or something might be true—by analogy with some known result—but it definitely needs a proof). Constructing a mathematical theory is thus guessing a web of ideas, and then progressively strengthening and modifying the web until it is logically unassailable. Before that point you don't have a theory. In fact, it is usually not assured at the beginning that you will be able to complete your construction as originally planned (otherwise, the theory would be uninteresting). Clearly, during your construction work, you should concentrate your efforts on the more uncertain links in your arguments. This is where your theory is most likely to fail, and you save time by knowing this early on. The easy and safe steps are left for later and are often handled in the final write-up by a dismissive sentence: "it is obvious that . . . ," "it is well-known that . . . ." Having secured the web of ideas that constitutes your theory, you still have to write things up, choosing an order of presentation, terminology, and notation, and hope that there will be no bad surprise in fixing the final details. Secondary considerations may play a big role in writing the final version of your article: relate your work to work by other mathematicians or state some intermediate result with more generality than strictly needed, so that it acquires independent interest. A good mathematician, who has spent a considerable amount of time in the primary work of elaborating a theory, may be more casual in the secondary elaboration that constitutes the final write-up. It is this casual attitude (I want to get this damn paper written and published, and forget about it) that will lead to mistakes, and typically those mistakes can be fixed without damaging the main results of the paper. We may say that our good mathematician, after spending a lot of time exploring a certain piece of mathematical landscape, writes a paper describing just one route in this landscape. If this route contains a forbidden shortcut, it is likely that another path can be found.

We have agreed that constructing a mathematical theory is the essence of mathematical work. Let me now try to outline some principles of strategy for such a construction. My approach will, by necessity, be informal. Keep in mind that the principles we know do not amount to a program that could be typed into a computer.

A first principle is that of *planning*. The construction of a mathematical theory starts with a plan, a web of more or less problematic ideas, which may well have to be profoundly modified later. Remember that when we discussed protein evolution by local mutations in chapter 20, we called this an effective but unintelligent tinkering process. Planning the construction of a mathematical theory, by contrast, may be called an *intelligent* process. Saying this just recognizes a difference between planned construction and tinkering, and gives the difference a name in agreement with common usage. (One can use the word *intelligent* without having first resolved the general metaphysical problem of defining intelligence. But beware that this use of the word has then no explanatory value).

Of course we now have to explain how to plan the construction of a mathematical theory, that is, how to set up a logically coherent web of mathematical ideas. Here I shall discuss some general principles—use of known facts and structural ideas, use of analogy—and finally I shall make some remarks about intuition.

Use of known mathematical facts includes application of known theorems in a way that may be easy and obvious. For instance, if you want to know the complex numbers z such that $z^2 - 3z + 1 = 0$, the fundamental theorem of algebra tells you that there are two such complex numbers, and a well-known formula gives them as $(3 - \sqrt{5})/2$ and $(3 + \sqrt{5})/2$ (which are real). Sometimes the application of the known theorems and formulas may be difficult and devious, and sometimes the use of a computer may be needed.[1] For some problems (like the simplification of algebraic expressions) obstinate tinkering is required and can be done by a computer program, with quite nontrivial results. Let me quote here some remarks pertaining to the computer software package Mathematica:

> The notion of transformation rules is a very general one. In fact, you can think of the whole of *Mathematica* as simply a system for applying a collection of transformation rules to many different kinds of expressions.

The general principle that *Mathematica* follows is simple to state. It takes any expression you input, and gets results by applying a succession of transformation rules, stopping when it knows no more transformation rules that can be applied.[2]

Use of structural ideas permeates contemporary mathematics. To take a simple example, suppose you encounter a set S such that, to elements $a, b \in S$, an element $a \times b \in S$ is associated. Then you *must* ask if the operation $\times$ is *associative* (i.e., if $(a \times b) \times c = a \times (b \times c)$) and if S with this operation is a *group*. If S is not a group, can one somehow extend S to a group? (Let me mention, without going into detail, that the obstinate will to introduce a group structure has led to an important area of study called *K-theory*, developed by a number of mathematicians after an original idea of Grothendieck.) To go back to an earlier discussion let me repeat that mathematical structures are a human invention. And in some cases (like measure theory) mathematicians do not agree on what is the natural structure to use. But structural considerations (including the use of categories and functors) are an essential feature in several branches of contemporary mathematics. In other branches, structural considerations do not seem to play such a prominent role. Yet, structural preoccupations of some kind are often present in the minds of mathematicians, even when not explicitly displayed. The structural approach to mathematics may be viewed as an ideological prejudice, but this prejudice has been remarkably fruitful, and one may say that it captures an important part of the obscure object of mathematical investigation, mathematical reality.

Analogy is a powerful tool for mathematical work, in particular during the planning stage of a theory. But unlike the use of known facts and structural ideas, analogy is not a safe guide. Here, from the fact that something is true in a certain situation, you want to guess that something related will be true in another situation that you think to be similar in some respect. For instance, knowing that there is an algorithm (Euclid) to divide an integer by another one (with a remainder), you can guess that something similar can be done with polynomials instead of integers. This sort of guesswork necessitates a good knowledge of mathematics, to have a good feel for what is similar and what isn't. The great virtue of analogy is that it can get you started in constructing a theory. But there is no guarantee that an analogy can be successfully pursued. Use of analogy is not a completely logical process, and this will delight some mathematicians, while irritating others. The latter will want to understand why two theories are similar, perhaps by finding a more general theory that contains both as special cases.

What about *mathematical intuition*? When we study a mathematical topic, we develop an intuition for it. We put in our memory a large number of facts that we can access readily and even unconsciously. Since part of our mathematical thinking is unconscious and part nonverbal, it is convenient to say that we proceed intuitively. This means that processes of mathematical thought are difficult to analyze. But it does not mean in my opinion that there is anything supranormal about mathematical intuition.

Speaking of supranormal leads me to mention a curious fact: mathematicians are more religious than most scientists. In fact the percentage of mathematicians who believe in God and afterlife is twice that of physicists.[3] I think this says that the relation of mathematicians to reality is—statistically—different from that of physicists. (Perhaps I should give my own view of the question: I am nonreligious, in a liberal sort of way. I fear equally the religious fanatics and the antireligious fanatics.)

Perhaps it is time to say a few words about *meaning* in mathematics. We have seen that the presentation of a mathematical theory in a technical article is somewhat removed from what the author had originally in mind. Intuitive ideas and nonverbal concepts have to be dressed up and couched in professional jargon. This may suggest that, hidden behind the formulas and jargon printed in professional journals, somewhere lies the true meaning of mathematics and that this is not of formal nature. In fact, during lectures (which are less formal than articles), the speaker will often explain what a theorem "really means." Why then not abandon the stilted formal language used for printed mathematics and explain the real meaning of what one is doing? To understand what is going on, remember that mathematics is a matter of knowledge, not of opinion. This is so because, since the Greeks, mathematics has had a solid basis of axioms and rules of deduction. From this basis, theories are developed. And from the theories, an intuition that goes ahead of the theories, recognizes analogies, and formulates conjectures. New results lead to new intuition that may in turn lead to a change in the logical structure of theories, with their axioms and definitions. But the intuitive meaning of mathematics is rooted in formalism. If one were to abandon the formalism and keep only the intuitive meaning, mathematics would soon be a matter of opinion rather than of knowledge. Its progress would then rapidly grind to a stop.

❖ 22 ❖

Mathematical Physics and Emergent Behavior

THE GREAT BOOK OF NATURE, according to Galileo, is written in mathematical language.[1] At least, one may say that students of the physical world, starting with Galileo, have taken up the task of transcribing the book into mathematics. And physicists are, in a sense, also mathematicians. But some physicists use very little math. Others, who call themselves *mathematical physicists*, use nontrivial mathematics in their studies of the great book. Newton, unquestionably, was a mathematical physicist. Einstein,[2] too, described himself as a mathematical physicist. Then there was a period in the mid-twentieth century when many physicists, Richard Feynman[3] among them, wanted to have nothing to do with mathematics. In fact, Feynman had a good knowledge of classical mathematics, and the *Feynman integral* that he introduced is a fundamental contribution to conceptual mathematics. The acquaintance of other physicists with mathematics was, however, often reduced to "a rudimentary knowledge of the Latin and Greek alphabets."[4] At the end of the twentieth century, mathematics came back in force in physics, with the popular *string theory*, which has led to important developments in pure mathematics but has so far made limited connection with the great book of nature. Currently, a number of papers that go under the heading of mathematical physics are made by people without much education in physics, and these contributions often have a somewhat doubtful scientific status. At the risk of belaboring the obvious, let me stress that the purpose of physics is not to prove "nontrivial physical theorems," it is to understand the great book of nature, using whatever method is found to work, and this may include the development of new mathematical theories.

The above remarks have no polemic intent: scientific colleagues who may read this chapter will know the complexity of the situation and have their own opinions on the subject. Other

119

readers are simply warned that "mathematical physics" may mean different things to different people. To me, mathematical physics has a unique character: Nature herself takes you by the hand and shows you the outline of mathematical theories that an unaided pure mathematician would not have seen. But many details remain hidden, and it is our task to bring them to light. I have found one aspect of this task particularly fascinating: making mathematical sense of emergent behavior of physical systems, as I shall explain in a moment.

Prominent in the history of physics is the discovery of the fundamental laws—those of classical mechanics and gravitation by Newton and Einstein, those of quantum mechanics by Heisenberg and Schrödinger.[5] From those fundamental laws one can now, in principle, understand *almost all* observed physical phenomena. Considerable efforts are currently under way to obtain a "theory of everything," allowing, in principle, to understand *all* observed physical phenomena. When such a theory has been obtained, it will be possible to compute every physical quantity, although perhaps with great difficulty and limited accuracy. It would then seem that the most interesting part of physics will be over, the rest being "just calculations." But this is not the case, because there are important conceptual problems in physics that go far beyond the discovery of the fundamental laws. The situation is, in fact, the same in various branches of mathematics. For instance, beyond the fundamental laws of arithmetic, there are important conceptual problems. Are there infinitely many primes? Are they distributed according to the prime number theorem? And so on.

Think now of understanding the properties of water, given that you know the fundamental laws of the mechanics of water molecules. You would, for instance, like to understand *phase transitions*: why, when you change its temperature, does water suddenly freeze to ice or boil to vapor? You would like to compute the *viscosity* of water (its resistance to deformation), and you would like to understand *turbulence*. (The fact that you can easily produce turbulence in your bathtub is no great help in understanding what it is.) The properties just mentioned are *emergent* properties. They are not properties of one water molecule or ten water molecules—they appear in the limit of infinitely many molecules. It is true that in the lab you always work with a finite amount of water, but the number of molecules in a liter of liquid is huge, and the properties of interest are (in first approximation) those of an infinite system.

I might be tempted to go into some details about phase transitions (equilibrium statistical mechanics), viscosity (nonequilibrium statistical mechanics), or turbulence. These, indeed, have been my professional fields of interest. But the technicalities involved are forbidding and would distract us from the purpose of this chapter: to discuss the relations between mathematics and mathematical physics in the study of emergent properties of many-particle systems. I would like to proceed with what I see as three important remarks on mathematical physics.

A first important remark is that *nature gives us mathematical hints*. In the case of water, the hint is to consider infinitely many water molecules in order to be able to discuss things like phase transitions or viscosity. But nature does not tell us everything, and it has taken the genius of Boltzmann and Gibbs,[6] and much further work, to understand how some problems, including phase transitions, can be analyzed in the framework of equilibrium statistical mechanics. This is a much simpler theory than the nonequilibrium statistical mechanics needed to study, for instance, viscosity. Nonequilibrium statistical mechanics requires that you consider the time evolution of an infinite system of molecules. This dynamical aspect, as we shall see in a moment, disappears in equilibrium statistical mechanics, from which time is absent.

A second important remark is that *mathematical physics deals with idealized systems*. We know that a water molecule is composed of oxygen and hydrogen nuclei surrounded by electrons and that the nuclei also have a composite structure. There is good reason to believe that these complications are not essential to understanding the freezing and boiling referred to above. A reasonable approach (in fact, the only feasible approach) is to study a variety of idealized systems. The simpler models can be analyzed more easily and in greater detail, and they may be mathematically more interesting. More elaborate models may be closer to physical reality and thus closer to the heart of physicists.

The third important remark is that *nature may hint at a theorem but does not state clearly under which condition it is true*. This is a complement to the first remark, and in a minute we shall see an example in the discussion of equilibrium statistical mechanics.

Equilibrium statistical mechanics is an *emergent theory*. It uses some concepts like *energy* that are already present in mechanics (classical or quantum) and other concepts that are new, like *equilibrium state* and *temperature*. I have to say that some states of matter are recognized by physicists as special and are called equilibrium states;[7] an example would be 1 kg of water at rest in a given volume V and at a given absolute temperature $T > 0$. The water consists of N molecules (corresponding to 1 kg), which can have various positions and velocities. (We choose here a classical rather than quantum description.) Classical equilibrium statistical mechanics describes the probability that the N particles occupy given positions and have given velocities. The water molecule H_2O may be variously oriented in space, and since this is an unwanted complication in the present discussion, we shall replace water by argon. The argon molecule is a single atom, which we may take to be spherically symmetric, and its position is thus given by the coordinates $\mathbf{x} = (x^1, x^2, x^3)$ of its center. (I shall now proceed with a couple of formulas that will clarify the subject for some readers; if they don't make sense to you, just read through rapidly.) Instead of the velocity $\mathbf{v}$, it is usual to consider the momentum $\mathbf{p} = m\mathbf{v} = (p^1, p^2, p^3)$ where m is the mass of an atom of argon. The energy of the N interacting argon atoms in the volume V is a function

$$E(\mathbf{x}_1, \ldots, \mathbf{x}_N, \mathbf{p}_1, \ldots, \mathbf{p}_N)$$

of the N positions (in the volume V) of the atoms and their N momenta. Classical equilibrium statistical mechanics gives the probability that each coordinate of the position is in an (infinitesimal) interval $(x_j^i, x_j^i + dx_j^i)$ and each component of the momentum is in an interval $(p_j^i, p_j^i + dp_j^i)$; this probability is

$$= Ce^{-E(\mathbf{x}_1, \ldots, \mathbf{x}_N, \mathbf{p}_1, \ldots, \mathbf{p}_N)/kT} \prod_{i=1}^{3} \prod_{j=1}^{N} dx_j^i \, dp_j^i,$$

where T is the absolute temperature, k is a universal constant (Boltzmann's constant), and the constant C is adjusted so that the integral over $\mathbf{x}_1, \ldots, \mathbf{x}_N$ in the volume V and $\mathbf{p}_1, \ldots, \mathbf{p}_N$ in $\mathbf{R}^3$ is 1.

For simplicity I have been discussing classical rather than quantum systems, and following Boltzmann and Gibbs, I have presented a certain probability measure that describes the equilibrium state of a system consisting of a large number N of particles. (This probability measure is known technically under the strange name of *canonical ensemble*.) Observe that the time evolution of our N particles has been forgotten. The idea (of Boltzmann, Gibbs, and others) is that there is an emergent behavior of a class of states of large systems (so-called equilibrium states) for which time evolution is irrelevant. The problem of justifying the emergent behavior of equilibrium states is most interesting but can be ignored if one so wishes; it is outside the scope of equilibrium statistical mechanics.

The founding fathers of equilibrium statistical mechanics were interested in the limit of large systems, and those systems have a very characteristic *extensive* behavior. Indeed, nature tells you that if, at a given temperature, you double the number of molecules and double the volume of the container (its shape doesn't matter too much), then also the energy of the equilibrium state should double. (More precisely, one should speak of the average energy in the equilibrium state and say that it doubles up to small corrections.) Nature thus tells you that, for a large system in equilibrium, there are *intensive* variables (temperature, pressure, . . .) and *extensive* variables (number of particles, volume, total energy, . . .) such that you can double the values of all extensive variables while keeping the values of the intensive variables fixed (up to small corrections). Clearly, there should be a theorem to justify this extensive or *thermodynamic* behavior, but nature does not tell us under what conditions the theorem holds. (This vagueness of nature's hints was our third important remark.) Does for instance a gas of stars show thermodynamic behavior? No! The globular star clusters observed by astronomers are not equilibrium states: they slowly shrink and evaporate. In fact, the attractive gravitational interaction between stars does not lead to thermodynamic behavior.

I have just sketched an example of emergent behavior (here, thermodynamic behavior) by which nature hints at a mathematical theory but leaves the details to be filled in by mathematical physicists. The study of other types of emergent behavior, as seen in nonequilibrium statistical mechanics or in hydrody-namic turbulence, is not less challenging.

The new mathematical structures uncovered by studies in mathematical physics may have considerable interest from a purely mathematical viewpoint and applications unrelated to physics. I shall now give an instance of this, with a brief techni-cal description that some readers may find a bit tough. Just go through it rapidly. You may be able, as in an earlier chapter, to appreciate the tune and way of singing, if not the detailed mean-ing of the song. I want to discuss the equilibrium statistical me-chanics of a system of *spins on a lattice*. Let us consider a finite box (drawn here as a piece of two-dimensional lattice) con-taining N spins $\sigma_1, \ldots, \sigma_N$:

```
+ + − + −
+ − − + −
− + + + +
− + − − +
− − + − −
```

Each spin in the box can take the value +1 or −1 (noted + or −), and a certain energy function $E(\sigma_1, \ldots, \sigma_N)$ is given. At temper-ature T, a spin configuration has then a probability

$$p_{\sigma_1 \cdots \sigma_N} = Ce^{-E(\sigma_1, \cdots, \sigma_N)/kT},$$

where the number C is adjusted so that the sum of all 2^N proba-bilities $p_{\sigma_1 \cdots \sigma_N}$ is 1. To observe thermodynamic behavior, we have to introduce so-called interactions, which allow the energy function for arbitrary large boxes to be computed (in a way that is invariant under lattice translations) and then take the limit of an infinite box:

```
·  ·   ·   ·   ·  ·
· + + − + − ·
· + − − + − ·
· − + + + + ·
· − + − − + ·
· − − + − − ·
·  ·   ·   ·   ·  ·
```

One can, in the limit, define a probability distribution for the system of infinitely many spins on the lattice, and this is called a *Gibbs state*. Gibbs states for spin systems on a lattice have a rich mathematical theory developed first by Dobrushin,[8] Lanford, myself, and then by many others, Sinai[9] among them. I looked in particular at one-dimensional systems:

$$\cdot \quad \cdot \quad + \quad - \quad + \quad + \quad + \quad - \quad - \quad \cdot \quad \cdot$$

and showed that for such systems there is only one Gibbs state and that it depends very nicely on the interaction (real-analytic dependence in some sense). This was not too astonishing: one-dimensional systems are physically expected to have no phase transitions (all this under suitable technical assumptions).

At this point the story of Gibbs states changes suddenly from mathematical physics to something else: Yasha Sinai proved the existence of *symbolic dynamics* for *Anosov diffeomorphisms*. This means that points of a suitable differentiable manifold M could be coded by sequences

$$\cdot \quad \cdot \quad + \quad - \quad + \quad + \quad + \quad - \quad - \quad \cdot \quad \cdot$$

corresponding to a one-dimensional spin system, in such a way that the action on M of a kind of differentiable map called an *Anosov diffeomorphism* corresponds to shifting all the symbols $\pm$ by one step to the left in the above sequence, that is, on a one-dimensional lattice. Sinai (and others, notably Bowen[10]) could then start the study of Gibbs states on manifolds. This idea had considerable mathematical developments[11] and returned from pure mathematics to physics in the study of *chaos*.[12] It has the great value of introducing an analytic tool (Gibbs states) into a geometric problem (diffeomorphisms). Note that one could, in principle, have proved the existence of symbolic dynamics without knowing equilibrium statistical mechanics, but Sinai had worked in statistical mechanics and was guided by his knowledge.

The above dry summary cannot convey the extraordinary experience that it was for me to be one of those involved in the development of a great mathematical idea, which started in a mathematical physics background and later returned to physics with chaos theory. It was also my great luck that those with whom I interacted directly were not only brilliant mathematicians but also unusually nice people. Indeed, there was a great period (a few years around 1970) when the Russians Dobrushin and Sinai, the Americans Lanford and Bowen, and myself exchanged ideas liberally, while new areas at the borderline of physics and mathematics were being opened.

❖ 23 ❖

The Beauty of Mathematics

Many of us find beauty in physical or biological artifacts of nature: a crystal of quartz, a flower, or a butterfly. And we find beauty in artifacts of man, like a perfectly shaped piece of pottery. And some of us also find beauty in mathematics.

Our sense of beauty belongs to human nature, and this is why we find beauty in a perfect human body, or human voice, or man-made pottery. But the sense of perfection, purity, and simplicity that we often associate with beauty also takes us away from the miseries of mankind: to flowers, crystals, the Gods, or God. We search for something beyond our ordinary human, biological, or physical world. Is there something beyond this world of uncertainties? There is: mathematics, which yields knowledge, not just opinions.

If I am going to speak of the beauty of mathematics, where logic rules, why mention the uncertainties of physics, biology, or theology? Simply because our human sense of beauty is not ruled by strict logic. Our sense of beauty may well guide us to a desire for inhuman logic. But this same sense of beauty remains very human and not particularly logical. Let me briefly mention in this respect that musical beauty is based on intervals that correspond to simple rational ratios between sound frequencies. But these rational ratios are messed up in our tempered scale system in a way that is acceptable to us humans mostly because we have a limited ability to distinguish between different sound frequencies. From an arithmetic point of view, the tempered scale system is a monstrosity: in our quest for musical beauty we have put convenience before logic.

We must be prepared to find that the perfection, purity, and simplicity that we love in mathematics is metaphorically related to a yearning for human perfection, purity, and simplicity. And this may explain why mathematicians often have a religious inclination. But we must also be prepared to find that our love of mathematics is not exempt from the usual human contradictions. What attracts many of us, frail humans, to mathematics is that it confronts the uncertainty and relativeness of human thought with the absolute certitude of mathematical truth. Only in mathematics can we check that a statement is correct by verifying each detail of its proof and then be absolutely sure of the conclusion, even if the proof is extremely long. Mathematics is the unique human endeavor where the use of a human natural language is, in principle, not necessary. Only here is no reference needed to our physical, biological, or psychological environment.

Among the various incentives that exist to do mathematics, one should mention the desire to be the best, the first, to become an important academician, and to win a million-dollar prize. I shall not dwell on these aspects, which are not really special to mathematics. What is important for many mathematicians is that they feel part of a select group of people sharing a common intellectual treasure. Something like this is also true for other human groups. But the mathematical community is special in agreeing on who belongs to it, being very international, interactive, relatively small (a few thousand creative mathematicians), and above all in consisting of a special group of people who have pushed intellectual achievement to its limit.

Mathematics is useful. It is the language of physics, and some aspects of mathematics are important in all the sciences and their applications and also in finance. But my personal experience is that good mathematicians are rarely pushed by a high sense of duty and achievement that would urge them to do something useful. In fact, some mathematicians prefer to think that their work is absolutely useless. (They may be wrong: number theory, traditionally viewed as beautiful and useless, has found applications in cryptography, with important financial and military aspects.) As to the applications of mathematics to physics and other sciences, I prefer in many cases to think of symbiosis. This symbiosis is a topic of great philosophical interest but is outside of mathematics proper, and I have limited myself to a brief discussion of the case of mathematical physics in the previous chapter.

Among useful things connected with mathematics, teaching is essential and is close to the heart of many mathematicians. Indeed, you may want to share your love for the beauty of mathematics, even if it is important for you that mathematics should remain useless! Teaching may be in the form of classes to students, seminar talks, or unorganized discussions. Mathematics has lived and flourished in a succession of places where it was taught and discussed: from Alexandria in antiquity, to Göttingen and Heidelberg during the nineteenth and twentieth centuries, and at many other places and times. It was my personal luck to be present in several places during the great periods where mathematics lived and was created,[1] and it is an unforgettable experience. But, in mathematics as in art, great periods do not last forever. And while the decline and fall of good places may occur in many different ways, politics often plays a decisive role: dictatorship at the country level or power games at the level of an individual institution.

I hope I have convinced you that the love of mathematical beauty is an essential reason why mathematicians do and teach mathematics. But can one say what makes mathematics beautiful? Let me propose one answer to this question: I think that *the beauty of mathematics lies in uncovering the hidden simplicity and complexity that coexist in the rigid logical framework that the subject imposes.*

Of course, the interplay and tension between simplicity and complexity are an element of art and beauty also outside of mathematics. Indeed, the beauty that we find in mathematics must be related to the beauty that our human nature sees elsewhere. And the fact that we are attracted by both simplicity and complexity, two contradictory concepts, befits our illogical human nature. But the remarkable thing here is that the shock of simplicity and complexity is intrinsic to mathematics; it is not a human construction. One may say that this is why mathematics is beautiful: it naturally embodies the simple and the complex that we are yearning for.

It is now time to be more concrete. Let me start with the reminder of two beautiful facts, historically old and important, both connected with the Pythagorean theorem. The first fact reveals unexpected simplicity: a triangle with sides of lengths 3, 4, 5 has a right angle opposite to the side of length 5. This premathematical observation strikingly points to hidden simplicity in the nature of things. The second fact is that the diagonal of a square with side 1 is irrational: $\sqrt{2} = 1.41421356\ldots$ cannot be written as the quotient of two integers. The proof of this fact shows that things are more complicated than one might have thought, and it forced the Greek mathematicians to accept the logical necessity of numbers that are not rational.

A general example of the interplay between simplicity and complexity is the fact that a short mathematical statement may need a very long proof. As a technical result this is a theorem of Gödel's that we have discussed in chapter 12. As a matter of fact, mathematicians know many theorems with a short statement (like Fermat's last theorem) and a very long proof.

There is an element of fashion in our judgment of mathematical beauty, as there is for artistic beauty. Bourbaki stressed a structural aspect of mathematics that constitutes an element of beauty for many modern mathematicians. But the aesthetic ideas of the ancient Greek were different and included a definite dislike of hubris. What would they think of mathematical proofs covering hundreds or thousands of pages? Clearly our intellectual landscape and our sense of beauty have changed over the centuries. Plato, Leonardo da Vinci, and Newton had different visions of the world, but each vision was unified, and man played a central role in it. Current science also strives to a unified view of the universe, but in this view, we humans appear as an insignificant accident. At the same time mathematical truth has acquired an even more fundamental role than physical reality. While the respective statuses of man and mathematics have been radically revised, the relations between the two partners have changed remarkably little since the Greeks. To get some understanding of this relation (one might say this beautiful relation) has been the object of this book.

And as we come to the end of our journey, let me say one more thing: it is while doing mathematical research that one truly comes to see the beauty of mathematics. It faces you in those moments when the underlying simplicity of a question appears and its meaningless complications can be forgotten. In those moments a piece of a colossal logical structure is illuminated, and some of the meaning hidden in the nature of things is finally revealed.[2]

❖ *Notes* ❖

Chapter 1: Scientific Thinking

1. D. Ruelle, "The obsessions of time," *Comm. Math. Phys.* **85** (1982), 3–5; "Is our mathematics natural? The case of equilibrium statistical mechanics," *Bull. Amer. Math. Soc. (N.S.)* **19** (1988), 259–268; "Henri Poincaré's 'Science et Méthode' " *Nature* **391**, (1998), 760; "Conversations on mathematics with a visitor from outer space" in *Mathematics: Frontiers and Perspectives*, ed. V. Arnold, M. Atiyah, P. Lax, and B. Mazur, Amer. Math. Soc., Providence, RI, 2000, 251–259; "Mathematical Platonism reconsidered," *Nieuw Arch. Wiskd. (5)* (2000), 30–33.

2. Isaac Newton (1643–1727) may be described as an English scholar or philosopher who is best remembered as a mathematician and theoretical physicist but also had quite different interests. His best biography remains that by R. S. Westfall (*Never at Rest: A Biography of Isaac Newton*, Cambridge University Press, Cambridge, 1980).

3. There was renewed public interest in the esoteric use of the Scriptures following the publication of Michael Drosnin's book *The Bible Code* (Simon and Shuster, New York, 1997). Drosnin's idea (unrelated to Newton's ideas) is that some sequences of equally spaced letters in the text of the Torah contain hidden meaningful information. This notion appeared to get support from some excellent mathematicians, but the reaction of scientists in general has been largely negative. See, for example, "The case against the codes" by Barry Simon (posted on the Internet).

Chapter 2: What Is Mathematics?

1. The Greek philosopher Pythagoras lived around 500 BC and remains a rather mysterious figure. Little is known of his mathematics and of his relation to the Pythagorean theorem.

2. The writings of the Greek philosopher Plato (427 BC–347 BC) remain surprisingly readable and exist in a nice compact English edition (*Plato: Complete Works*, ed. J. M. Cooper and D. S. Hutchinson, Hackett Publishing, Indianapolis, 1997). Of course, when you read Plato you should allow yourself to disagree with him: his logic is sometimes

questionable by modern standards, and his political ideas may seem to us occasionally close to Fascism. But mostly one has the delightful impression of conversing with a very intelligent, open-minded, and pleasant man.

3. Euclid lived around 300 BC in Alexandria, Egypt. The thirteen books of his *Elements* constitute the most important monument of Greek mathematics that is left to us.

4. The German mathematician David Hilbert (1862–1943) was a towering figure of mathematics. His version of Euclidean geometry was presented in the book *Grundlagen der Geometrie* in 1899. Among other things, Hilbert is famous for twenty-three problems (unsolved at the time) that he proposed to the mathematical community at the International Congress of Mathematicians in Paris in 1900. In 1930 he expressed his optimism on the power of mathematics in the statement,

Wir müssen wissen, wir werden wissen.
(We must know, we shall know.)

Gödel's paper of 1931 showed, however, that there are limitations to what one can know.

5. The Austrian-born mathematician and logician Kurt Gödel (1906–1998) proved some stunning results on the logical structure of mathematics. His incompleteness theorems published in 1931 show that in any (not too simple) axiomatic mathematical system there are propositions that cannot be proved or disproved in the system. In particular, the consistency of the axioms cannot be proved.

6. J. P. Serre, *Cours d'arithmétique*, Presses Universitaires de France, Paris, 1970; English trans.: *A course in Arithmetic*, Springer, Berlin, 1973. Jean-Pierre Serre (1926–) is a French mathematician.

7. S. Smale, "Differentiable dynamical systems," Bull. Amer. Math. Soc. **73** (1967), 747–817. Stephen Smale (1930–) is an American mathematician.

CHAPTER 3: THE ERLANGEN PROGRAM

1. The German mathematician Felix Klein (1849–1925) made fundamental contributions to geometry.

2. *Real and Complex Numbers*

The distance (in some unit) between points O and X on a line is a positive number d (or 0 if X coincides with O). Having fixed O, the position of X is determined if we give d a sign: + or − according to whether X

is to the right or to the left of O. We say that $+ d$ or $- d$ is a real number. Call it x: it may be positive, negative, or zero. So, a real number x determines exactly the position of a point on the line (once you have chosen O, a unit of length, and which are the right and the left side of O).

left ———————————————————————→ right
O X

A complex number is an expression like $x + iy$, where x and y are real numbers and i is a new symbol. It is assumed that i multiplied by itself (i.e. i^2) is -1. Saying that $x + iy, = 0$ means that both x and y are 0. You can add, subtract, and multiply complex numbers (for multiplication, use $i^2 = -1$). If $x + iy \neq 0$, you can also divide by $x + iy$, in fact

$$\frac{1}{x + iy} = \frac{x}{x^2 + y^2} - \frac{iy}{x^2 + y^2}.$$

Now draw two lines in the plane, meeting perpendicularly at O. We call these lines the x-axis Ox and the y-axis Oy, as below.

From a point Z draw perpendiculars ZX to Ox and ZY to Oy. Call x the distance from X to O with a $+$ or $-$ sign depending on whether X is right or left of O, and call y the distance from Y to O with a $+$ or $-$ sign depending on whether Y is above or below O. In this way we have a correspondence between the point Z of the plane and the complex number $z = x + iy$. In other words we can think of complex numbers as points in the plane (one then speaks of the *complex plane*).

CHAPTER 4: MATHEMATICS AND IDEOLOGIES

1. We have seen in note 2 of chapter 3 how the position of a point Z in the plane can be coded by two real numbers x, y. Similarly, a point in three dimensions can be described by three numbers. The systematic use of this idea is due to the French philosopher and mathematician

René Descartes (1596–1650). This has permitted mathematicians to apply algebra to geometry, a development that has been fundamental for geometry and mathematics in general.

2. A. Vershik, "Admission to the mathematics faculty in Russia in the 1970s and 1980s," *Mathematical Intelligencer* 16 (1994), 4–5; A. Shen, "Entrance examinations to the Mekh-mat," (*Math. Intelligencer*) **16** (1994), 6–10. These articles are reprinted with a mathematical study of entrance examination problems by I. Vardi and other contributions in a volume entitled *You Failed Your Math Test, Comrade Einstein* (ed. M. Shifman, World Scientific, Singapore, 2005).

3. As my wife puts it, there are fewer bastards and fewer frauds among mathematicians than in the general population, but maybe also fewer amusing people!

CHAPTER 5: THE UNITY OF MATHEMATICS

1. An example of this is the *Riemann Hypothesis*, a famous conjecture on the distribution of large primes formulated by the German mathematician Bernhard Riemann (1826–1866). By the diversity and depth of his work, Riemann was perhaps the greatest mathematician of all time, though he died before reaching the age of forty. Proving the Riemann Hypothesis is the eighth of the twenty-three problems proposed by Hilbert in 1900.

2. The Swiss-born mathematician Leonhard Euler (1707–1783) proved this formula in 1734. At that time Euler lived in Saint Petersburg, where he died.

3. The German philosopher and mathematician Gottfried Wilhelm Leibniz (1646–1716) developed a version of the infinitesimal calculus. It is questionable how independent his results were with respect to those of Newton, but his notation is still used today.

4. The German mathematician Georg Cantor (1845–1918) did foundational work in set theory.

5. The English mathematician and logician Alan Turing (1912–1954) also left deep conceptual contributions in other fields. He will appear again in chapter 15.

6. The French mathematician Alexander Grothendieck (1928–) will appear again in Chapters 6 and 7.

7. The Belgian-born mathematician Pierre Deligne (1944–) worked in France and now works in the United States.

8. The Séminaire Bourbaki survives and is still very active. Three times per year there is a meeting in Paris, where five lectures are given, carefully presenting some topics of current interest. The text of the lectures, written in advance, is distributed to the audience. The Séminaire Bourbaki plays a significant role in the dissemination of (some) new mathematical ideas.

CHAPTER 6: A GLIMPSE INTO ALGEBRAIC GEOMETRY AND ARITHMETIC

1. The work of the French mathematician Henri Poincaré (1854–1912) covers many topics, and his books on the philosophy of science remain very current and readable.

2. This is a special case of Bézout's theorem.

3. While remembered mostly for his work in number theory, the French mathematician Pierre Fermat (1601–1665) was also a lawyer and counselor at the parliament of Toulouse.

4. The proof of Fermat's last theorem results from contributions by a number of mathematicians, but the decisive final (and most difficult) step was taken by the British mathematician Andrew Wiles (1953–), who currently works in the United States.

CHAPTER 7: A TRIP TO NANCY WITH ALEXANDER GROTHENDIECK

1. Paul Montel (1876–1975) belongs to a French tradition that associates excellent mathematics and a long life. Jacques Hadamard (1865–1963) and Henri Cartan (1904–) are in the same tradition.

2. Motchane appears in Vercors' *La bataille du silence* (Les Editions de Minuit, Paris, 1992). This beautiful book tells about some men and women who defied the French authorities and their Nazi masters during World War II by publishing books illegally, at a time when it could have cost them their lives.

3. The American theoretical physicist J. Robert Oppenheimer (1904–1967) played an essential role in the development of the atomic bomb.

4. The French mathematician René Thom (1923–2002) was an independent mind. He was not a member of Bourbaki, and he spent quite a bit of time on his catastrophe theory and questions of philosophy.

But he may in the end be remembered mostly for important contributions to geometry.

5. Some information on Grothendieck's background is contained in two related papers by P. Cartier, one in French ("Grothendieck et les motifs," IHES preprint, 2000) and one in English ("A mad day's work, . . .," trans. Roger Cook, Bull. Amer. Math. Soc. **38** (2001), 389–408). Cartier has made a considerable and fairly successful effort to save Grothendieck from premature burial, but I am skeptical about his "psychoanalytic" interpretations of Grothendieck. Another source of interest is A. Herreman ("Découvrir et transmettre," IHES preprint, 2000), which discusses the "coup de poing en pleine gueule"; see below. There is also an excellent article by Allyn Jackson, "*Comme appelé du néant*—as if summoned from the void: The life of Alexandre Grothendiek" (Notices Amer. Math. Soc., **51** (2004), I, 1038–1056; II, 1196–1212). And I have used my own memories, supported by personal archives, on the period discussed.

For a mathematical discussion of Grothendieck's work, see J. Dieudonné ("De l'analyse fonctionnelle aux fondements de la géométrie algébrique," in *The Grothendieck Festschrift*, I, Prog. Math. 86, Birkhäuser, Boston, 1990, 1–14). Let me quote Dieudonné's concluding lines on the work of Grothendieck in algebraic geometry: "It is out of the question to summarize these six thousand pages. There are few examples in mathematics of such a monumental and fertile theory, built in such a short time and due essentially to a single man."

6. The French mathematician Jean Dieudonné (1906–1992) was one of the main figures of Bourbaki. He was an early member of the IHES.

7. Or at least very different. Let me try to justify this statement. Many successful scientists leave us a story of their life. These autobiographies typically contain interesting personal and historical information, amusing anecdotes, and suggestions that the author had interests in life other than just science (like music, sex, administration, . . .). The story culminates with a handshake between the great man of science and some other great man, president, king, or perhaps pope. If you read Grothendieck's *Récoltes et semailles*, you may not like it, but you sense a very different personality.

8. The French theoretical physicist Louis Michel (1923–1999) was one of the early members of the IHES.

9. "Vous êtes un fieffé menteur, Monsieur Motchane." This is quite strong language and upset a number of people who had sympathies for Grothendieck.

10. A. Grothendieck's *Récoltes et semailles* (1985–86) is available on the Internet in various forms and translations, together with further material. What is available changes with time, and the reader is invited to check for herself or himself.

11. The Crafoord Prize, to be shared with Pierre Deligne. The Crafoord Prize is given by the Swedish Academy of Sciences for work in various non-Nobel disciplines.

12. See note 5.

13. Pierre-Gilles de Gennes (1932–) is a French theoretical physicist.

CHAPTER 8: STRUCTURES

1. We have just introduced two set-theoretical concepts: *subsets* and *maps* (or *functions*). Here are two more definitions. The *intersection* of two sets S and T is the set, denoted by $S \cap T$, consisting of those elements that belong to both S and T. The *union* of S and T is the set, denoted by $S \cup T$, consisting of all those elements that belong to S, T, or both. So $\{a,b\} \cap \{a,c\} = \{a\}$, $\{a,b\} \cap \{c\} = \varnothing$ (the empty set), and $\{a,b\} \cup \{a,c\} = \{a,b,c\}$. One can also define the intersection and union of more than two sets (general families of sets, possibly infinite).

2. The Polish-born American mathematician Samuel Eilenberg (1913–1998) and the American Saunders Mac Lane (1909–2005) collaborated in the 1940s and 50s.

3. The Hungarian-born mathematician Paul Erdös (1913–1996), with his lasting attachment to his mother, his addiction to amphetamines, and other unusual traits of character, may appear as a somewhat extreme personality. It is remarkable that the very special environment provided by mathematics allowed this personality to flourish.

4. M. Aigner and G. M. Ziegler, *Proofs from The Book*, Springer, Berlin, 1998 (3rd ed. in 2004). Incidentally, if you look at theorem 1 of chapter 8, "In any configuration of n points in the plane, not all on a line, there is a line which contains exactly two of the points," you are tempted to use the methods of projective geometry to get a proof. You will find in *Proofs from The Book* an explanation of why this doesn't work!

5. See chapter 2, note 2.

6. Lest I be misunderstood, let me stress that I do not adhere to a literary view of science that has been popular in certain circles (namely, that a scientific text, like any other piece of literature, is just a reflection of the socioeconomic conditions under which it was produced and has

to be studied as such). I believe that the literary approach misjudges the scientific content of scientific texts and that literary criticism is a limited way to explore the relations of the human mind with the science it produces.

CHAPTER 9: THE COMPUTER AND THE BRAIN

1. The Hungarian-born American scientist John (earlier Johann) von Neumann (1903–1957) has been thought by some to have been Stanley Kubrick's model for Dr. Strangelove. (The Hungarian-born American physicist Edward Teller (1908–2003) has also been proposed.)

2. J. von Neumann, *The Computer and the Brain*, Silliman Memorial Lectures Vol. 36, Yale University Press, New Haven, CT, 1958.

3. The Greek scientist Archimedes of Syracuse (287–212 BC) is remembered for his engineering and physical ideas but mostly for his contributions to mathematics. His calculations of surfaces and volumes anticipate the infinitesimal calculus of Newton and Leibniz, and he is considered one of the greatest mathematicians of all time.

4. There is a common view that thinking is equivalent to speaking. For example, Plato writes (in *Sophist*): "Aren't thought and speech the same, except that what we call thought is speech that occurs without the voice, inside the soul in conversation with itself?" (see *Plato's Complete Works*, ed. J. M. Cooper and D. S. Hutchinson, Hackett Publishing, Indianapolis, 1997, p. 287). The practice of mathematical thinking shows the importance of nonverbal elements, in particular, visual elements.

5. Computers can make random errors, so-called glitches. The removal of such errors (by repetition, checks) has been studied. But with current technology, the error level is so low that it is not relevant to the present discussion.

6. See "Conversations on mathematics with a visitor from outer space," *Mathematics: Frontiers and Perspectives*, ed. V. Arnold, M. Atiyah, P. Lax, and B. Mazur, Amer. Math. Soc., Providence, RI, 2000, 251–259.

CHAPTER 10: MATHEMATICAL TEXTS

1. A detailed analysis is given by Reviel Netz (*The Shaping of Deduction in Greek Mathematics: A Study in Cognitive History*, Ideas in Context, 51, Cambridge University Press, Cambridge, 1999).

2. You may want to exercise your skills on the following problem. Take three mutually perpendicular axes Ox, Oy, Oz in space. Consider the solid circular cylinder C_x with axis Ox and radius R, and similarly for C_y, C_z. These three cylinders (with equal radii) intersect in a solid S limited by curved faces. Question: what does S look like? How many faces does it have, what are their shapes, and how are they put together? Combining visual intuition and reasoning, you can get the correct answer, but this is somewhat painful. A drawing on a sheet of paper makes things much easier. (Draw the intersections of two cylinders at a time.)

3. Different languages offer different poetical possibilities because rhythm, grammar, vocabulary, word similarities, and clusters of meaning are different. For instance, German is strongly accented, and this is powerfully used in Goethe's verses:

> Wer reitet so spät durch Nacht und Wind?
> Es ist der Vater mit seinem Kind.

But the weak accent in French can be used with great subtlety, as when Apollinaire writes:

> Le colchique couleur de cerne et de lilas
> Y fleurit tes yeux sont comme cette fleur-là.

The different factors of form, meaning, and word associations sometimes conspire to yield the miraculous thing that is a great poem. I am not convinced by the attempts I have seen at translating poetry in verses: it is too much to expect the same miracle to happen twice in different languages. I treasure, however, the help provided by honest translations in prose for poems (like those of Saint John of the Cross in Spanish) that would otherwise elude me because my knowledge of the original language is limited or nonexistent.

4. Most mathematicians now type their own manuscripts on a laptop or a computer terminal, using some appropriate software like TeX. A formula written in TeX is tolerably close to a sentence in English; in fact, (*) is typed in as

```
$${U - A\over M - A} : {U - B\over M - B} =
{M - A\over V - A} : {M - B\over V - B}$$
```

Incidentally, TeX has been a great invention for blind mathematicians, who can read the above formula (a linear arrangement of a limited variety of symbols) more easily than the original (*).

CHAPTER 11: HONORS

1. Giordano Bruno (1548–1600) was an Italian philosopher and heretic. To him and to countless others who have suffered and are suffering for speaking when the authorities of the time wanted them to be silent, we owe what freedom of speech we enjoy.

2. Some people provide supremely great performance and show in sports and are richly rewarded by the public for their act. There is nothing wrong with that. What is wrong is that money domination encourages the use of drugs and cheating, and defeats the justification of sports as health activities. There is also nothing wrong with generously rewarding great scientific achievements. In fact, I am all in favor of that. But giving money an excessive role has its dangers, and caution is needed. Cheating (presentation of fabricated results, among other things) has become a problem in medicine, biology, and physics and there are signs that the corrupting effects of money may not forever spare mathematics.

CHAPTER 12: INFINITY: THE SMOKE SCREEN OF THE GODS

1. The German mathematician and theoretical physicist Ernst Zermelo (1871–1953) made fundamental contributions to set theory.

2. The German-born mathematician Adolf Fraenkel (1891–1965) moved to Jerusalem in 1929.

3. *Encyclopedic Dictionary of Mathematics* (2nd ed., 4 vol., MIT Press, Cambridge, Mass., 1987) is a translation from the Japanese (3rd ed., K. Itô, Mathematical Society of Japan, Tokyo, 1985. It is remarkable how much of significant twentieth-century mathematics could be presented in this compendium.

4. As an example of paradox arising in *naive* set theory, let me mention Russell's paradox, which goes as follows. Say that x is a set of the first kind if it does not contain itself as an element ($\neg\, x \in x$) and of the second kind if it contains itself as an element ($x \in x$). A set must be of the first or the second kind and cannot be both. Call X the set of all sets of the first kind. If X is of the first kind, X does not belong to X, that is, does not belong to the set of sets of the first kind. This is a contradiction because X is of the first kind. If X is of the second kind, X belongs to X, that is, belongs to the set of sets of the first kind. This is a contradiction because X is of the second kind. What this means is

that introducing notions like "the set of all sets" is inviting trouble and is not allowed in serious *axiomatic* set theory.

The English logician and philosopher Bertrand Russell (1872–1970) is also remembered for his pacifist political positions.

5. This has been proved under the assumption that the set of axioms one starts with is sufficiently rich to develop the theory of natural integers.

6. At the same time, Gödel also proved the following result. Starting with a set of axioms sufficiently rich to develop the theory of natural integers, it is not possible to prove the consistency of this system of axioms by utilizing arguments formalizable in the theory developed from these axioms. There are consistency proofs for some mathematical theories, but such proofs use stronger theories.

7. The American logician Alonzo Church (1903–1995) proposed in 1936 a precise definition of *effectively calculable functions*. This proposal is known as *Church's thesis*, and one version of it (the Church-Turing thesis) is that an effectively calculable function is one that can be calculated by a Turing machine, which is a simple computer (finite automaton) with an unlimited memory as described by Turing. One may require that upon each input the machine gives an answer in finite time (this corresponds to computing a *general recursive function*) or allow the machine to not always give an answer (this corresponds to computing a *partial recursive function*). Most mathematicians want to be allowed to work with more general functions than those that are effectively calculable.

8. To be precise: the maximum length of a proof of a statement of length L is not a general recursive function of L. This statement is not sensitive to the precise definition of length of statement or length of proof. For some details, see, for instance, Yu. I. Manin, *A Course in Mathematical Logic*, trans. Neal Koblitz, Grad. Texts in Math. **53**, Springer, New York, 1977, Section VII.8.

CHAPTER 13: FOUNDATIONS

1. Let me describe a *group* in typical mathematical jargon: neither the formal language of logicians nor the baby talk of popular science writers.

Let G be a nonempty set, and when $a,b \in G$, let $c \in G$ be given, called the product of a and b, and write $c = ab$. We call G, equipped with this

141

product, a group (or we say that the product defines a group structure on the set G) if the following properties (called axioms of a group) hold:

   (i) associativity: a(bc) = (ab)c;

   (ii) existence of unit element: there exists e ∈ G such that for every a ∈ G, ea = ae = a;

   (iii) existence of inverses: for every a ∈ G there exists x ∈ G such that ax = xa = e.

Note that the unit element *e* is unique. If *ab* = *ba*, the group is said to be *commutative*. If *G,G′* are groups with unit elements *e,e′* and *f* is a function from *G* to *G′* such that *f* (*ab*) = *f* (*a*) *f* (*b*), then *f* is called a *morphism* *G* → *G′*, and the subset *H* of elements *x* ∈ *G* such that *f* (*x*) = *e′* is called a *normal subgroup H* of *G*. If the only normal subgroups *H* of *G* are {*e*} and *G*, then *G* is called a *simple group*.

My reason for going into all these technical details is that I can now make a statement: group structure is important. Specifically, if you find a group structure in the problem you are studying, it will help you. And it should be automatic for you to ask if the group is commutative or not and to look for its normal subgroups. As an example, the transformations associated with various geometries in chapter 3 form *groups* of transformations (Euclidean, affine, or projective). Groups appear in a useful way in the practice of mathematics; this is why they are natural objects, not because the definition of group structure is relatively simple.

2. We have prudently introduced the complex numbers in chapter 3 (see note 2) and pictured them as points in a plane (the complex plane). We now denote the complex plane by **C** and recall that **C** is a field (from chapter 6, we know that complex numbers can be added, multiplied, and divided). *Analytic* (or *holomorphic*) functions of a complex variable are functions *f* defined on a subset *D* of **C**, with values in **C** and such that, for *z* in *D* and |*z* − *z*₀| sufficiently small, one can express *f* (*z*) as an infinite sum

$$f(z) = \sum_{n=0}^{\infty} a_n \, (z - z_0)^n ,$$

where the a_n are complex numbers. Analytic functions have remarkable properties. In particular, if *f* is analytic in *D*, there are usually larger subsets *D̃* of **C** such that *f* extends (uniquely) to an analytic function *f̃* on *D̃* (this extension is called *analytic continuation*).

3. Riemann saw that the distribution of prime numbers could be related to properties of a function now known as the *Riemann zeta function*. Following Riemann's idea, Hadamard and de la Vallée-Poussin proved a result known as the *prime number theorem*. This says that the number of primes $\leq n$ tends to infinity like $n/\ln n$, where $\ln n$ is the logarithm of n. The Riemann Hypothesis is a conjectured property of the zeta function that would lead to a refinement of the prime number theorem.

Jacques Hadamard (1865–1963) was a French mathematician, and Charles de la Vallée-Poussin (1866–1962) was a Belgian mathematician.

4. There are other important axiom systems than those for set theory, notably Peano arithmetic (PA), which axiomatizes the theory of integers. But PA is much weaker than ZFC. So, while PA is interesting for logicians, it is not much used by "normal" mathematicians.

5. The Axiom of Choice (C) says the following:

If a set X contains subsets A_λ, indexed by $\lambda \in \Lambda$, and no A_λ is the empty set $\varnothing$, then we can choose x_λ in A_λ for each $\lambda \in \Lambda$ (i.e., there is a function $f : \Lambda \to X$ such that $f(\lambda) \in A_\lambda$ for each $\lambda \in \Lambda$).

Once you understand what it means, you will probably find (C) intuitively acceptable, but note that $x_\lambda = f(\lambda)$ is in no way explicitly constructed.

6. Stefan Banach (1892–1945) was a Polish mathematician, and Alfred Tarski (1902–1983) was a Polish-born logician. The Banach-Tarski "paradox" is the following fact that can be proved using the Axiom of Choice:

It is possible to cut a solid sphere (in three-dimensional space) into finitely many pieces and, after moving these pieces around (by three-dimensional rotations and translations), to reassemble them into two solid spheres of the same size as the original one.

The number of pieces can be taken to be five. This may seem absurd because the volume of the pieces adds up to the volume of one sphere at the beginning and to twice this volume at the end. There is, however, no real paradox because one cannot speak of the volume of the pieces that are moved around: these pieces are *nonmeasurable*. When the Axiom of Choice is used to produce sets, these sets are usually nonmeasurable. Nonmeasurability is a nuisance, but the current consensus of mathematicians is that they like to have the Axiom of Choice at their disposal, even at the cost of having to be a bit careful about measurability of the sets that they manipulate.

7. Not only is (C) consistent with ZF, it is independent of ZF as shown by P. Cohen: if ZF is consistent, there is a consistent axiom system including ZF but such that the Axiom of Choice does not hold.

The American mathematician Paul Cohen (1934–) is especially known for his work on the axiomatic foundations of set theory, using a technique called *forcing*.

8. A case in point is the theory of Banach spaces, where an important result, the Hahn-Banach theorem, requires the Axiom of Choice for its proof. Use of Hahn-Banach permits a nicer general theory of Banach spaces, and since the theory of Banach spaces is fairly useful in applications, a puritanical attitude prohibiting use of the Axiom of Choice is not very welcome here.

9. Finite simple groups are simple groups (see note 1) that are finite sets. These algebraic objects can be classified, that is, listed: the list is infinite but quite explicit. While the experts consider that the classification work is now complete, the publication of the proofs needed to support the classification is still going on and is remarkable by its length, many thousands of pages of hard technical mathematics. (See, for instance, R. Solomon, "On finite simple groups and their classifications," *Notices Amer. Math. Soc.* **42** (1995), 231–239; M. Aschbacher, "The status of the classification of the finite simple groups" *Notices Amer. Math. Soc.* **51** (2004), 736–740.)

10. We have already met polynomials on several occasions, in particular in chapter 6. Consider finitely many *variables* $z_1, \ldots, z_v$, a *monomial* in these variables is a product

$$c z_1^{n_1} \cdots z_v^{n_v},$$

where c is a *coefficient* and $n_1, \ldots, n_v$ are natural integers. So, a monomial is obtained by raising the variables $z_1, \ldots, z_v$ to some powers $n_1, \ldots, n_v$, multiplying the $z_j^{n_j}$, and multiplying the product by the coefficient c. A polynomial $p(z_1, \ldots, z_v)$ is a finite sum of monomials as above. For instance,

$$p(x, y) = c + c' x + c'y$$

is a polynomial in the two variables x, y (with coefficients c, c', c'), and

$$p(x, y, z) = x^n + y^n - z^n$$

is a polynomial in three variables. In classical algebraic geometry, the coefficients are complex numbers and so are the variables.

Consider now a polynomial $P(x_1, \ldots, x_\mu, y_1, \ldots, y_\nu)$ in the $\mu + \nu$ variables $x_1, \ldots, x_\mu, y_1, \ldots, y_\nu$, where the coefficients are integers (positive, negative, or zero). We shall associate with this polynomial P a set S of points $\langle a_1, \ldots, a_\mu \rangle$ where $a_1, \ldots, a_\mu$ are natural integers, that is, elements of $\mathbf{N} = \{0, 1, 2, 3, \ldots\}$. In other words, the points of S will be sequences $\langle a_1, \ldots, a_\mu \rangle \in \mathbf{N}^\mu$. The set S is defined to consist of those $\langle a_1, \ldots, a_\mu \rangle$ for which there exist natural integers $b_1, \ldots, b_\nu$ such that

$$P(a_1, \ldots, a_\mu, b_1, \ldots, b_\nu) = 0.$$

Whenever there is a polynomial $P(x_1, \ldots, x_\mu, y_1, \ldots, y_\nu)$ such that the subset S of $\mathbf{N}^\mu$ can be defined as just indicated, we say that S is a *Diophantine* set.

Theorem. A subset S of $\mathbf{N}^\mu$ is Diophantine if and only if it is recursively enumerable.

This results from the work of a number of mathematical logicians, the proof being completed in 1970 by Yuri Matijasevič. Remember from chapter 12 that a set S is recursively enumerable if there is an algorithm that systematically lists all its elements. But it may not be possible to list the elements not in S. In this case we have very little control on S, and we may not know whether S is empty or not. So, the above theorem gives a negative solution to Hilbert's tenth problem, which asked for an algorithm to tell, for any polynomial $P(x_1, \ldots, x_\mu)$ with integer coefficients, if there are integers $a_1, \ldots, a_\mu$ such that $P(a_1, \ldots, a_\mu) = 0$. In fact, there can be no such algorithm. Nevertheless, the unsolvability of Hilbert's tenth also has positive consequences. For instance, we know, by the above theorem, that the set of all primes (which is a subset of $\mathbf{N}$) is Diophantine.

See M. Davis, "Hilbert's tenth problem is unsolvable," *Amer. Math. Monthly* **80** (1973), 233–269; M. Davis, Yu. Matijasevič, and J. Robinson, "Hilbert's tenth problem: Diophantine equations: Positive aspects of a negative solution" in *Mathematical Developments Arising from Hilbert Problems (Northern Illinois Univ., De Kalb, Ill., 1994)*, Proc. Sympos. in Pure Math. **28** (1974), 323–378.]

The Greek mathematician Diophantus of Alexandria probably lived in the third century AD and left a collection of problems known as *Arithmetica* (algebra and number theory).

11. Let the region D in the complex plane $\mathbf{C}$ consist of the complex numbers $z = x + iy$ (x and y real) such that $x > 1$. The Riemann zeta function is defined on D by the infinite sum

$$\zeta(z) = \sum_{n=1}^{\infty} \frac{1}{n^z}.$$

One can show that ζ is an analytic function in D and that it has a unique analytic continuation (again called ζ) to the complex plane $\mathbf{C}$, minus the point 1. Consider the subset R of $\mathbf{C}$ consisting of the complex numbers $z = x + iy$ such that $1/2 < x < 1$. One formulation of the Riemann Hypothesis is that ζ does not vanish in the "forbidden" region R (that is, $\zeta(z) \neq 0$ if $z \in R$). It is known that $\zeta(z)$ vanishes at $z = -2$, $-4, -6, \ldots$, and also at an infinity of points $z = 1/2 + iy$; the usual formulation of the Riemann Hypothesis is that there are no other zeros.

12. S. Shelah, "Logical dreams," *Bull. Amer. Math. Soc. (N.S.)* **40** (2003), 203–228.

CHAPTER 14: STRUCTURES AND CONCEPT CREATION

1. See chapter 13, note 1.
2. See chapter 2, note 4.
3. See chapter 12, in particular note 8.
4. See chapter 13, note 2.
5. The statement is the *maximum modulus principle* for analytic functions. Let me give a precise statement without speaking of boundary. If $f(z)$ is analytic in the domain $D = \{z: |z - z_0| < R\}$ (disk of radius R centered at z_0) and if there exists $a \in D$ with $f(a) \neq f(z_0)$, then there exists $b \in D$ with $|f(b)| > |f(z_0)|$, that is, the modulus of $f(z)$ cannot be maximum at the center of a disk in which $f(z)$ is analytic.

6. The concept of a compact set belongs to topology, and I cannot pretend to give you a good idea of topology in this note if you have never studied the subject before. But it is easy enough to give the basic definitions, just to show how absurdly simple they are. (We shall use the concepts of subset, map, intersection, and union, for which you may see chapter 8, note 1; the words *family,* and *subfamily* (of sets) may here be understood as set of sets and subset of a set of sets.)

A topology on a set X is a family of subsets of X, called open sets, such that the following axioms are satisfied:

(1) X and the empty set ∅ are open sets;

(2) the intersection of two open sets is an open set;

(3) the union of any family of open sets is an open set.

Suppose we have topologies on both the set X and the set Y, and let f be a map from X to Y. For a subset V of Y we denote by $f^{-1}V$ the set of points $x \in X$ such that $fx \in V$. With this notation, the map f is said to be *continuous* if, whenever V is open in Y, then $f^{-1}V$ is open in X.

One says that subsets O_i of X (a possibly infinite family) form a covering of X if the union of all O_i is X. A space X with a topology is said to be *compact* if, for any covering of X by open subsets O_i, there is a finite subfamily of sets O_i that already form a covering of X.

Suppose that X and Y have topologies and that f is a continuous map from X to Y such that $fX = Y$ (for every point $y \in Y$ there is some $x \in X$ such that $fx = y$). Then, if X is compact, it follows that Y is also compact.

Having read this concise description of topology, you might say, "I, too, am a mathematician," and start writing your own axioms, definitions, and theorems. What is not assured is that they will be as significant for mathematics as the conceptual skeleton of topology that I have just described.

7. See note 6.

8. Abstract measure theory starts by giving a measure (or mass) $m(X)$ to certain subsets of a space M. The theory of Radon measures assumes M to be a compact topological space and starts by defining an integral (or average value) $m(A)$ for continuous functions A on M. Abstract measure theory is more general. The theory of Radon measures is a special case and has, therefore, more theorems: it is a *richer* theory.

9. See M. R. Garey and D. S. Johnson, *Computers and Intractability*, Freeman, New York, 1979.

10. This is what I did in the paper "Conversations on mathematics with a visitor from outer space" in *Mathematics: Frontiers and Perspectives*, ed. V. Arnold, M. Atiyah, P. Lax, and B. Mazur, Amer. Math. Soc., Providence, RI, 2000, 251–259.

11. This statement has, in fact, to be corrected. The slow, blind work of evolution has generated mechanisms (in the immune system and, of course, the nervous system) that produce relatively fast and intelligent responses.

12. This is actually the title of the "first part" of the treatise, but Bourbaki didn't go much beyond that.

13. This is actually J.-P. Serre quoting A. Grothendieck, in a letter of February 8, 1986. The letter is in response to Grothendieck after the latter had sent *Récoltes et semailles* to Serre. In this very interesting letter, Serre acknowledges the power of Grothendieck's approach but expresses the opinion that it does not work in all parts of mathematics. See *Correspondance Grothendieck-Serre*, ed. P. Colmez and J.-P. Serre, Documents Mathématiques 2, Société Mathématique de France, Paris, 2001.

CHAPTER 15: TURING'S APPLE

1. Not only is the number π irrational, but it is, in fact, *transcendental*; that is, it cannot satisfy the equation

$$a_n\pi^n + a_{n-1}\pi^{n-1} + \cdots + a_1\pi + a_0 = 0$$

when $a_0, a_1, \ldots, a_{n-1}, a_n$ are integers (positive, negative, or zero). This was proved in 1882 by the German mathematician Ferdinand von Lindemann (1852–1939).

2. The Danish mathematician Harald Bohr (1887–1951) is remembered for his theory of almost periodic functions. He was also the brother of the physicist Niels Bohr (1885–1962) and a member of the 1908 Danish Olympic soccer team.

3. Ioan James, "Autism in mathematicians," *Math. Intelligencer* **25** (2003), 62–65.

4. Constance Reid, *Hilbert*, Springer, Berlin, 1970.

5. The German-born American mathematician Richard Courant (1888–1972) was a student and later a collaborator of Hilberts.

6. Andrew Hodges, *Alan Turing: The Enigma*, Simon & Schuster, New York, 1983.

7. "Can machines think?" To test this, Turing proposed that an interrogator would ask questions of a person and a machine locked in a different room. The person and the machine would type answers that might be lies (the machine pretending that it is a person). Could the interrogator find out which was the person and which was the machine? That is the *Turing test*: an imitation game in which the machine must pass as a person. If person and machine cannot be told apart, it is hard to say that the machine cannot think. Interestingly, in presenting the imitation game, Turing used a woman and a man instead of person and a machine.

8. Playing at home with dangerous chemicals was probably more common and less discouraged in the early 1950s than it is now. At the time that Turing did his experiments with cyanide, I was a teenager and had a little lab in the basement where I experimented with arsenic (As_2O_3), phosphorous, and other poisonous, inflammable, explosive, corrosive, or foul-smelling substances.

9. Frank Olver is now professor emeritus in the Math Department of the University of Maryland. I am indebted to him for telling me of the time when he knew Turing at the National Physical Laboratory, in England in the late 1940s.

10. A few of the mathematicians that I have known were openly gay, but I would not say that being gay is common in the profession. And, if you must know, I am not gay, and I did not have a nervous breakdown. As for being bald, I shall admit to having a receding hairline. Well, to be quite honest, a permanently receded hairline.

CHAPTER 16: MATHEMATICAL INVENTION: PSYCHOLOGY AND AESTHETICS

1. H. Poincaré, "L'invention mathématique" (Mathematical creation), *Science et méthode*, Ernest Flammarion, Paris, 1908, chapter 3; English trans. in *Science and Method*, Dover, New York, 1952.

2. J. Hadamard, *The Psychology of Invention in the Mathematical Field*, Princeton University Press, Princeton, NJ, 1945; enlarged 1949 edition reprinted by Dover, New York, 1954.

3. Einstein's letter is reproduced as Appendix II in Hadamard's book; see note 2.

4. There are exceptions. Poincaré's philosophical writings (see, for instance, note 1) are good literature. Interestingly, young Henri Poincaré had started to write a novel. From what we know of the novel, it is no great loss to us that he abandoned this project. But his early literary preoccupations were clearly an asset when Poincaré later started writing on the philosophy of science.

5. The implicit function theorem plays a foundational role in differential geometry (the study of differential manifolds). See, for instance, S. Lang, *Differential Manifolds*, Addison-Wesley, Reading, MA, 1972.

6. An example is the proof of persistence of hyperbolic sets; see M. W. Hirsch and C. C. Pugh, "Stable manifolds and hyperbolic sets"

in *Global Analysis (Berkeley, Calif.; 1968), Proc. Sympos. in Pure Math.* **14,** Amer. Math. Soc., Providence, RI, 1970, 132–163.

7. There are actually several ergodic theorems: a pointwise ergodic theorem (Birkhoff) and a mean ergodic theorem (von Neumann) appeared in 1932; other ergodic theorems have followed. These theorems permit the definition of "time averages" and play a foundational role in ergodic theory. (See, for instance, P. Billingsley, *Ergodic Theory and Information*, John Wiley & Sons, New York, 1965.)

CHAPTER 17: THE CIRCLE THEOREM AND AN INFINITE-DIMENSIONAL LABYRINTH

1. See T. D. Lee and C. N. Yang, "Statistical theory of equations of state and phase transitions, II: Lattice gas and Ising model," *Physical Rev.* (2) **87** (1952), 410–419, and also T. Asano, "Theorems on the partition functions of the Heisenberg ferromagnets," *J. Phys. Soc. Japan* **29** (1970), 350–359. I have long been fascinated with the Lee-Yang circle theorem (see D. Ruelle, "Extension of the Lee-Yang circle theorem," *Phys. Rev. Lett.* **26** (1971), 303–304), and I think that there are still mysteries to be unveiled in this area.

2. The fundamental theorem of algebra is a theorem of analysis more than of algebra. It says that for a polynomial $P(z) = \sum_{j=0}^{m} a_j z^j$, where the a_j are complex numbers and $a_m = 1$, there exist complex numbers $c_1, \ldots, c_m$ such that $P(z) = \prod_{j=1}^{m} (z - c_j)$.

CHAPTER 18: MISTAKE!

1. The Chinese mathematician Shiing-shen Chern (1911–2004) spent a good part of his career in the United States.

2. The references are H. Hopf, "Über die Abbildungen der dreidimensionalen-Sphäre auf die Kugel-fläche," *Math. Ann.* **104,** (1931) 637–665; "Über die Abbildungen von Sphären auf Sphären niedrigerer Dimension," *Fund. Math.* **25** (1935), 427–440.

3. An algorithm solves a certain type of problem upon presentation of suitable data. For instance, the problem can be *is this integer a prime?* and the integer for which the question is asked is the data. The data have a certain *length*, which is here the number of digits of the integer. It is of obvious interest to know how fast an algorithm is, that is, how

long it will take to solve a given problem. For instance, for a polynomial-time algorithm, the execution time is bounded by a polynomial in the length of the data. A problem is considered *tractable* if it has a polynomial-time algorithm. Remarkably, there is a polynomial-time algorithm for primality testing (i.e., to find out if an integer is a prime or not), but no polynomial-time algorithm is known to find the prime factors of an integer that is not a prime. (The tractability of primality testing was proved in 2002 by M. Agrawal, N. Kayal, and N. Saxena.) For some problems, if one can guess an answer, this answer can be checked in polynomial time, and there is a class of such problems that are in some sense equivalent (NP-complete class; see note 9 of chapter 14). A big open question is whether NP-complete problems can actually be solved in polynomial time. It is generally assumed that this is not the case, but there is no proof.

4. The Poincaré conjecture (1904) characterizes the three-dimensional sphere among three-dimensional manifolds. After many other attempts, it seems that Grigori Perelman finally proved the Poincaré conjecture in 2002.

5. A. Jaffe and F. Quinn, "Theoretical mathematics' toward a cultural synthesis of mathematics and theoretical physics," *Bull. Amer. Math. Soc. N.S.* **29** (1993), 1–13; M. Atiyah et al., "Responses," *Bull. Amer. Math. Soc. N.S.* **30** (1994), 178–207.

6. So-called strange attractors; see, for instance, J.-P. Eckmann and D. Ruelle, "Ergodic theory of chaos and strange attractors," *Rev. Modern Phys.* **57** (1985), 617–656.

7. Suppose that the surface of the sphere is cut into "countries" (there are no seas), that each country is connected (not composed of disjoint pieces), and we want to color the countries so that countries with a common boundary have different colors (we allow same-color countries that have only a finite number of boundary points in common). How many colors do we need? K. Appel and W. Haken in 1977 published a computer-aided proof finding that four colors are sufficient.

8. The American mathematical physicist Oscar E. Lanford (1940–) made several important contributions to statistical mechanics. His computer-aided proof is unpublished. I discussed it earlier in "Mathematical Platonism reconsidered"; see chapter 1, note 1.

9. See M. Aschbacher, "The status of the classification of the finite simple groups," *Notices Amer. Math. Soc.* **51** (2004), 736–740.

10. If you put spherical marbles in a cubic container, the maximum density that can be achieved in the limit of a large container is known as the *close-packing density* (for spheres). This density is reasonably easy to guess, but to prove the guess seems extremely hard. See T. Hales "The status of the Kepler conjecture," *Math. Intelligencer* **16** (1994), 47–58, and B. Casselman, "The difficulties of kissing in three dimensions," *Notices Amer. Math. Soc.* **51** (2004), 884–885.

11. See note 5 above.

Chapter 19: The Smile of *Mona Lisa*

1. An obvious remark is that, perhaps, in spite of what I think and claim, the unnamed seminar speaker actually said *antisymmetric*. The *anti-Semitic* that I heard would then be a production of my unconscious, not his. I shall not argue the reasons I think as I do but note that, clearly, somebody's unconscious was at work. And, for the purposes of the present discussion, it is not essential to know whose it was.

2. The original German title is *Eine Kindheitserinnerung des Leonardo da Vinci*. To the art historian Meyer Schapiro we owe a fundamental and extremely readable study of Freud's book ("Leonardo and Freud," *Journal of the History of Ideas* **17** (1956), 147–179). I have read the Kindheitserinnerung in a bilingual French-German version (*Un souvenir d'enfance de Léonard de Vinci*, Gallimard, Paris, 1995), with a long preface by the psychoanalyst J.-B. Pontalis, who uses M. Schapiro's study. Freud's *Leonardo* gives a possible and very interesting interpretation of the personality of the great artist and is excellent reading. Just keep an alert and critical mind! And, by the way, another book that I found great reading is Freud's *Moses and Monotheism* (*Der Mann Moses und die monotheistische Religion*, Verlag Allert de Lange, Amsterdam, 1939).

3. Some people believe that Leonardo was homosexual. Perhaps you would prefer to think that he had an incredibly romantic love affair with a stunningly beautiful Florentine noblewoman, breathless and tragic, but you can't accept the notion that he had no sex life at all! Still, Freud could guess better than most people what was going on in a person's mind, and he may well have been right here.

4. These are Leonardo's quotations in Freud's book.

5. See chapter 1, note 2.

6. I am here following in part J. Laplanche and J.-B. Pontalis, *Vocabulaire de la psychanalyse*, Presses Universitaires de France, Paris, 1981.

Chapter 20: Tinkering and the Construction of Mathematical Theories

1. How strong the bias is may be described by a temperature: strong bias = low temperature. The standard picture here is to find an energy minimum rather than a maximum of interest. High temperature corresponds to a random walk that jumps around a lot, paying only minor attention to lowering the energy. When using a computer, a good strategy, called *simulated annealing*, is to start at high temperature (visiting a lot of ground without getting stuck and settling in a wide region of low energy). Then the temperature is progressively lowered (to refine the selection of a low-energy value).

2. F. Jacob "Evolution and tinkering," *Science* **196** (1977), 1161–1166. The French biologist François Jacob (1920–) is known for his groundbreaking work on regulatory activities in bacteria.

3. Evolution could, of course, have done all kinds of different things. I like to think that it might have produced vertebrates with six legs instead of four, so that one pair of legs could have more easily been freed to obtain wings or arms. Well-known imaginary creatures could then have become real: dragons (with four legs and two wings), centaurs (with four legs and two arms), and angels (with two legs, two arms, and two wings). (See D. Ruelle, "Here be no dragons," *Nature* **411** (2001), 27].

4. Aharon Kantorovich, *Scientific Discovery, Logic and Tinkering*, State Unversity of New York Press, Albany, 1993.

Chapter 21: The Strategy of Mathematical Invention

1. For instance, D. Zeilberger has given a (Maple) computer program to prove identities involving hypergeometric functions ("A fast algorithm for proving terminating hypergeometric identities" *Discrete Math.* **80** (1990), 207–211).

2. See Section 1.4.1. in S. Wolfram, *The Mathematica Book*, Cambridge University Press, Cambridge, 1996.

3. See the correspondence by E. J. Larson and L. Witham in "Leading scientists still reject God," *Nature* **394** (1998), 313. This shows low figures (14.3 percent for mathematicians, 7.5 percent for physicists) for religious belief among members of the National Academy of Science

(USA). Higher percentages available on the Internet for other samples show the same factor of 2 for religious inclination of mathematicians over physicists.

CHAPTER 22: MATHEMATICAL PHYSICS AND EMERGENT BEHAVIOR

1. The Italian mathematician, astronomer, and physicist Galileo Galilei (1564–1642) is one of the founders of modern science. His freedom of thinking got him in trouble with the Catholic Church of the time. It is interesting to speculate with which authorities he would have gotten in trouble if he had lived in the present period. Galileo insisted that philosophy should be studied in the great book of the world, written by nature, not in the texts of the Greek philosopher Aristotle (384–322 BC). In the *Saggiatore* (1623) is a famous quotation: "La filosofia è scritta in questo grandissimo libro che continuamente ci sta aperto innanzi a gli occhi (io dico l'universo). . . . Egli è scritto in lingua matematica." (Philosophy is written in this very great book that is there constantly open to our eyes (I mean the universe). . . . It is written in mathematical language.)

2. The German-American Albert Einstein (1879–1955) was probably the greatest physicist of the twentieth century.

3. The American theoretical physicist Richard Feynman (1918–1988) reworked in depth several aspects of quantum mechanics.

4. Here is what the Swiss mathematical physicist Res Jost (1918–1990) wrote: "In the thirties, under the demoralizing influence of quantum-theoretic perturbation theory, the mathematics required of a theoretical physicist was reduced to a rudimentary knowledge of the Latin and Greek alphabets" (quoted by R. F. Streater and A. S. Wightman, *PCT, spin and statistics, and all that*, W. A. Benjamin, New York, 1964).

5. Modern quantum mechanics dates from its mathematical formulation in 1925 by the German Werner Heisenberg (1901–1976) and, in a different form, in 1926 by the Austrian Erwin Schrödinger (1887–1961).

6. An essential role in the conceptual foundation of statistical mechanics was played by the Austrian Ludwig Boltzmann (1844–1906) and the American J. Willard Gibbs (1839–1903).

7. Note that physics always involves an essential nonmathematical element: the operational identification of "things" in nature for which one will try to find a mathematical description. To have an equilibrium

state of water, one has to let water rest for a suitable time, check that the water does not move, build a thermometer and check that the temperature does not depend on place or time, and so on.

8. The Russian Roland L. Dobrushin (1929–1995) was an outstanding probabilist who took interest in equilibrium statistical mechanics and obtained deep results in that area.

9. The Russian mathematician Yakov G. Sinai (1935–) made fundamental contributions to the ergodic theory of dynamical systems and to statistical mechanics.

10. The American mathematician Robert E. Bowen (1947–1978), known as Rufus Bowen, made essential contributions to the theory of smooth dynamical systems. (He told me that he chose the name Rufus because he disliked being called Bob.) He was the opposite of an agitated genius: when he explained some mathematical problem in his slow, quiet voice, you forgot everything except the question that he was describing, with absolute clarity. He was one of the very best mathematicians of his age when he unexpectedly died of brain hemorrhage.

11. Some of the ideas mentioned here are discussed in the following technical books: R. Bowen, *Equilibrium States and the Ergodic Theory of Anosov Diffeomorphisms*, Lecture Notes in Math. **470**, Springer, Berlin, 1975; D. Ruelle, *Thermodynamic Formalism: The Mathematical Structures of Classical Equilibrium Statistical Mechanics*, Addison-Wesley, Reading, MA, 1978; and W. Parry and M. Pollicott, *Zeta Functions and the Periodic Orbit Structure of Hyperbolic Dynamics*, Astérisque **187–188**, Soc. Math. de France, Paris, 1990.

12. See, for instance, my nontechnical book *Chance and Chaos*, Princeton University Press, Princeton, 1991.

CHAPTER 23: THE BEAUTY OF MATHEMATICS

1. As far as dynamical systems are concerned, I made several visits to Berkeley in the 1960s and '70s during Steve Smale's great "hyperbolic" period, then to the Instituto Nacional de Matemática Pura e Aplicada in Rio de Janeiro when Jacob Palis and Ricardo Mañé flourished. As to mathematical physics, I was in Zurich (Eidgenössische Technische Hochschule Zurich) with Res Jost in the early '60s, then at the Institute for Advanced Study in Princeton at the time of C.-N. Yang

and Freeman Dyson. I also benefitted from the statistical mechanics activity around Joel Lebowitz at Yeshiva University and later Rutgers. And of course I was immersed in the constant mathematics and mathematical physics activity at the IHES in Bures sur Yvette for several decades in the second half of the twentieth century.

2. And this is the end of your toil, O you, the patient reader of notes! We are leaving the academy and its disputes behind us. We can now breathe some fresh air and allow ourselves to be again, for a while, $\alpha\gamma\varepsilon\omega\mu\varepsilon\tau\rho\eta\tau o\iota$, that is, geometrically inept, or nonmathematicians.

Index